A Practical Handbook of Preparative HPLC

Don Wellings is the Chief Scientific Officer for Chromatide Ltd, a company specializing in contract purification and consultancy services. Don has been performing preparative chromatographic separations since his PhD where he was routinely running 10 cm diameter columns more than 20 years ago. The wealth of knowledge accumulated over the successive years has been as broad as it is deep.

Prior to setting up Chromatide with an ex-colleague from Avecia, he was Technology Manager for Special Projects at Polymer Laboratories, where he was intimately involved in the design and development of new polymeric stationary phases for reversed phase, normal phase, ion-exchange, affinity and chiral HPLC. Previously Don was Technology Manager for Separation Sciences and Solid Phase Organic Chemistry at Avecia. During this period he developed an expertise applying molecular modeling to the design of chiral ligands for preparative chromatography.

In the late 1990s he was involved in the installation and commissioning of process scale HPLC at Zeneca Pharmaceuticals. During 18 years with CRB, ICI, Zeneca and Avecia he was instrumental in establishing the technique of preparative HPLC within the company and served 11 years as the secretary of the company's Process Scale Chromatography Group.

A Practical Handbook of Preparative HPLC

Dr. Donald A. Wellings

AMSTERDAM • BOSTON • HEIDELBERG • LONDON • NEW YORK •
OXFORD • PARIS • SAN DIEGO • SINGAPORE • SYDNEY • TOKYO

Elsevier
The Boulevard, Langford Lane, Kidlington, Oxford, OX5 1GB, UK
Radarweg 29, PO Box 211, 1000 AE Amsterdam, The Netherlands

British Library Cataloguing in Publication Data
Wellings, Donald A.
A practical handbook of preparative HPLC
1. High performance liquid chromatography
I. Title
543.8′4

Library of Congress Cataloging-in-Publication Data
A catalog record for this book is available from the Library of Congress

ISBN 13: 978-1-8-56-17466-4
ISBN 10: 1-8-56-17466-2

Typeset by Charon Tec Ltd, Chennai, India
www.charontec.com

Printed and bound by CPI Group (UK) Ltd, Croydon, CR0 4YY

Transferred to Digital Print 2011

Contents

Preface

This text is intended to be a guide for both the novice to preparative HPLC, and as an aid to the chemical engineer planning to introduce this 'black art' into the industrial environment.

The first question to ask is '*What is preparative?*' To many, the isolation of a few grams of an extremely potent molecule may be considered as largescale. In some instances 50 g of a vaccine will supply the annual market for a particular disease state. In more traditional drug therapies a few tonne may be more typical.

The second question to be answered is '*What is HPLC?*' This abbreviation is often derived from the term 'High Performance Liquid Chromatography', though the term 'High Pressure Liquid Chromatography' is often preferred since high performance can also be achieved at low pressure. Just to confuse the issue, is this the pressure created by the resistance to liquid flow through the column or, the pressure at which the column is packed?

To help you to decide whether you have picked up the correct book let's be practical. This book will describe particles packed into columns. These stationary phases are rigid porous media typically in the range of 5–30 μm in size and the columns you are interested in are predominantly pre-packed at 2000–6000 psi or you are going to self-pack your own dynamic axial compression columns at 50–100 bar.

Too many chromatographic texts dwell heavily on a theoretical and mathematical complexity that bears little relevance to what you actually need to do in order to practice preparative HPLC. Hopefully this book will describe how to practically go about a preparative separation. It is designed to guide the reader through the choice of equipment and chromatographic modes with minimal fuss and with reference to only relevant formulae. Much of the 'black art' will be removed by the hints and tips of a practitioner with over 20 year's experience in many modes of chromatographic separation.

Finally, if you know what dynamic axial compression (DAC) is then you have the correct book so read on.

Foreword

Don Wellings asked me to write a foreword to his book and I am honoured and glad to do so. I have known Don for more than fifteen years and I place him among the top prep chromatographers in the world today, alongside people like Gregor Mann and Jules Dingenen.

Having been involved from the start in the creation and the establishment of the Kromasil silica-based media business, during many years as General Manager, I have experienced the impressive development of preparative HPLC over the last twenty years. The technique is now as important to learn as other standard operations, such as distillation and crystallization. It is often the only way of achieving sufficiently high purity of biotech products.

Preparative HPLC plays a large role in the education programs for chemical engineers and will do so even more in the future. I have to admit that I myself have not read any book about preparative HPLC except this one – the reason being that when I graduated in 1965 there were few, if any, books available on the subject. I am convinced, however, that this book is an ideal one for use at the universities, or for anybody interested in Preparative HPLC.

For regulatory people not directly involved in the technical process, this book gives a very good guidance in how to deal with validation issues like GMP. If you know little about DAC, and if you are not experienced in optimizing HPLC processes by utilizing positive self displacement and avoiding tag along, this book will be of high value.

The book is very nicely written and can very well defend its place among any other book you may read, whether in a laboratory setting, or even during your vacation perhaps in a sailing boat moored in a quiet natural harbour, or in a comfortable chair under a shady tree in an English garden.

Hans Liliedahl
Founder
Triple Moose Technologies
Västra gatan 51B
SE-44231 Kungälv

Abbreviations

APIs	active pharmaceutical ingredients
cGMP	current good manufacturing practice
DAC	dynamic axial compression
DQ	design qualification
de	diastereoisomeric excess
ee	enantiomeric excess
FATs	factory acceptance tests
FDA	Federal Drug Agency
FTEs	full-time equivalents
HETP	height equivalent to a theoretical plate
HPLC	high performance liquid chromatography
IQ	installation qualification
IUPAC	International Union for Pure and Applied Chemists
MSDSs	material safety data sheets
OQ	operational qualification
PQ	performance qualification
PIs	process instructions
PRs	process records

SATs	site acceptance tests
SEM	scanning electron micrograph
SMB	simulated moving bed
SOPs	standard operating procedures
URS	user requirement specification

1

The history and development of preparative HPLC

Chapter One

Chromatography can be defined as the separation of mixtures by distribution between two or more immiscible phases. Some of these immiscible phases can be gas–liquid, gas–solid, liquid–liquid, liquid–solid, gas–liquid–solid and liquid–liquid–solid. Strictly speaking, a simple liquid–liquid extraction is in fact a chromatographic process. Similarly, distillation is a chromatographic process that involves separation of liquids by condensation of their respective vapours at different points in a column.

Most will remember the school science project of placing an ink blot in the centre of a filter paper and following this by dripping methylated spirits on to the ink. Watching in fascination as concentric circles of various pigments develop is probably the first and sometimes last experience of a chromatographic separation many will encounter. Like too many of our observations the essence of this experiment is to demonstrate that black ink is made up of several different pigments and the underlying process, in this case chromatography, is dismissed with blatant disregard.

A Colourful origin!

Chromatography was originally developed to isolate coloured pigments from plants. Hence, from Greek origins we get chromato, 'colour' and graph, 'to record'.

Fortunately for us, some very clever scientists have seen the 'wood for the trees' and have taken these simple observations and developed them into complex, highly efficient, methods of purification.

The invention of chromatography was rightly accredited to Mikhail Tswett in 1902[1.1] for his detailed study of the selective adsorption of leaf pigments on various adsorbents, though somewhat unwittingly, the first demonstrations of preparative chromatography probably stem back to 'bleaching' of paraffin by passage through a carbon bed in the 1860s.

The saviour of many a frustrated chemist!

Mikhail Tswett was neither chemist or chemical engineer. In fact, he was a botanist researching in the isolation of plant pigments.

The first column based separations performed in a true industrial setting can be better demonstrated by the purification of petroleum on Fuller's earth in the 1920s. The 1950s marked the development of simulated moving bed (SMB) chromatography for the separation of sucrose and fructose in the sugar industry. However, these separations are limited low to medium pressure chromatography since the columns could be packed and operated in place. The high pressure generated by the small particles used as stationary phases in HPLC dictates the use of specialist hardware. The columns are generally machined from a solid ingot in order to avoid the flaws that can be observed in welded columns. The weight of the thick walled columns normally limits the scale at which columns can be manually handled so it is unusual to find pre-packed columns with a greater than 10 cm diameter. Scaling beyond this requires fixed hardware and it can be said that the first true high pressure based preparative chromatographic separation did not arrive until the 1980s following the invention of dynamic axial compression (DAC) based columns.

DAC, invented by Couillard[1,2] led to a dramatic change in philosophy. The column packing operation could now be developed and carried out at the point of application. Subsequently, the scale of preparative separations would now only be limited by the column design. The DAC concept involves the constant compression of the packed column bed during a separation, allowing for the concomitant removal of column dead space formed as the bed height reduces during operation. The reduction of the bed height under flow is usually attributed to a more regular rearrangement of the stationary phase particles within the column or due to degradation and dissolution of the stationary phase itself.

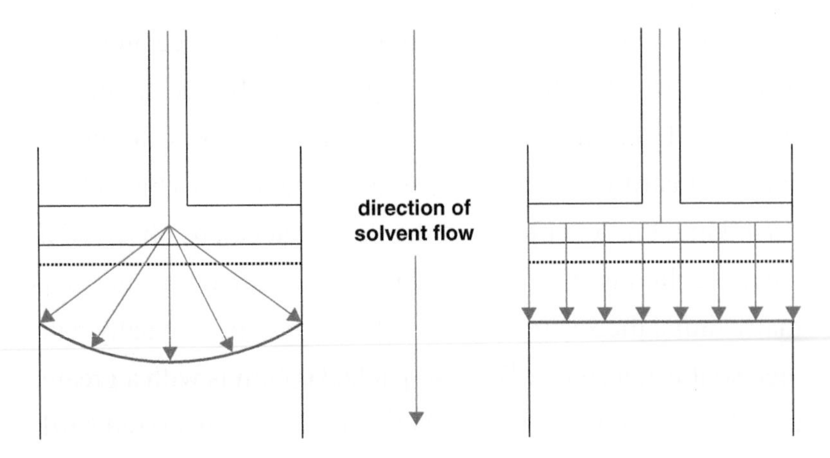

Figure 1.1

Probably the most important issue that had to be overcome as the scale of operation increased was the engineering of even flow and sample distribution over larger column diameters. There are many ways of distributing the sample at the inlet, and similarly collecting the eluate but the basic principle is to deliver solvent to all points across the column diameter simultaneously. The flow through a column end fitting is shown schematically in Figure 1.1, where the left hand diagram demonstrates poor distribution resulting in a convex solvent front, shown in red, and the right hand side shows the optimum sample delivery.

Various distribution plates have been designed using anything from simple engineering logic[1.3,1.4,1.5] to computational fluid dynamics (CFD)[1.6]. Layouts vary from complex multi-layered plates[1.7] to single discs, but the most common approach

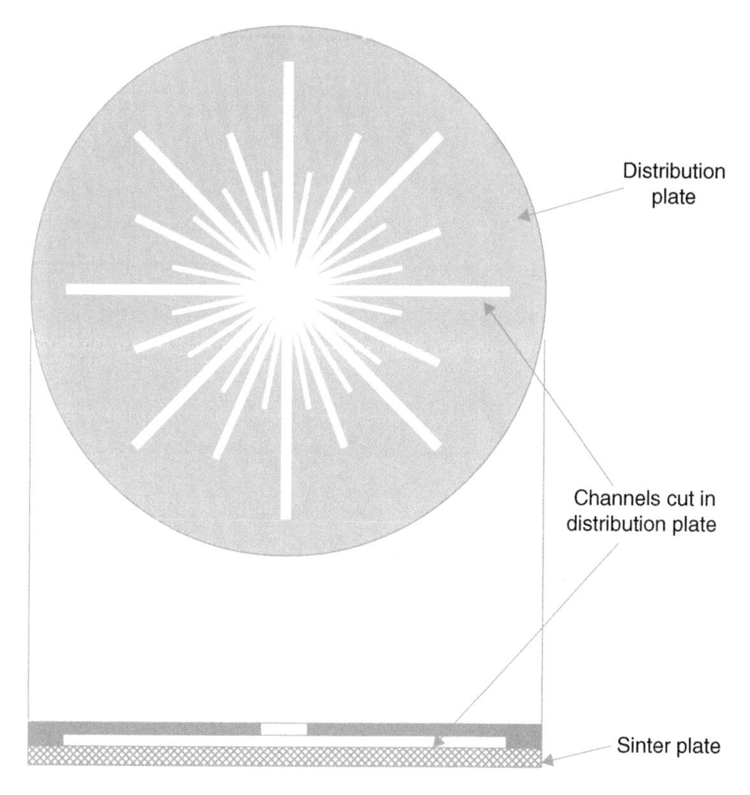

Figure 1.2 Schematic of a typical distribution plate

is to use a star type distribution plate represented in schematic form in Figure 1.2 and shown photographically in Figure 1.3. The strategically placed and sized holes and channels allow for a near simultaneous release of eluate over the surface area of the column. The sinter plate, in contact with the distribution plate on one side and stationary phase on the other, improves the dispersion further.

Figure 1.3 Courtesy of Jerome Theobald, NovaSep SAS

The increase in scale of preparative HPLC, brought about predominantly by the invention of DAC, resulted in a proportionate demand for high quality stationary phases. A move from the rather crude irregular silica based media used for normal

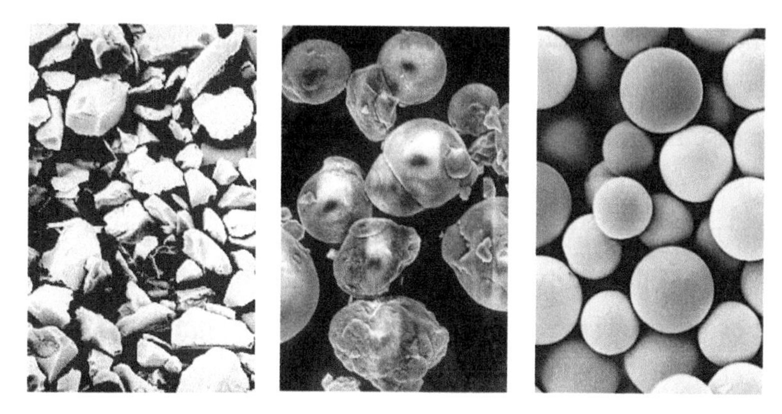

Figure 1.4 Courtesy of Per Jageland, Eka Chemicals and Gregor Mann, Schering AG

phase chromatography towards spherical particles was now inevitable. Figure 1.4 shows the dramatic changes that have taken place in moving from the irregular particles of yesteryear to the highly developed spherical particles now available. The DAC of irregular materials leads to a mechanical degradation resulting in the generation of fines, which ultimately results in product contamination and blockage of the column frits. A search for the optimum spherical silica based stationary phase with the enhanced mechanical stability required for process scale DAC has fuelled a whole new market for the media.

Even though DAC soon became established as the method of choice, it took a further fifteen years before stationary phases with uniform particle size and pore size, with the prerequisite mechanical stability, started to appear.

The various modes of operation, including normal phase, reversed phase ion exchange and chiral chromatography, will be discussed later. However, whatever the mode of separation, it is essential to have an understanding of the precise source of the media. Nowadays, though a number of suppliers can deliver high quality silica it is important to note that the supplier is not always the manufacturer. Some suppliers subcontract the core silica manufacture and will carry out surface modifications in-house to provide a range of normal phase, reversed phase and chiral stationary phases. This must be considered when working to current Good Manufacturing Practice (cGMP) and should be included as part of the vendor qualification process if a long-term, robust supply chain is required. The number of suppliers that manufacture and modify stationary phases can be counted on one hand and the silica based market is currently dominated by one major supplier.

The growing popularity of reversed phase chromatography in particular has prompted polymer manufacturers to investigate the use of polymeric media for this mode of operation. Macroporous copolymers of styrene and divinylbenzene have similar properties to silica based stationary phases bonded with alkyl chains. However, the absence of leachables and stability at high pH can offer advantages under certain circumstances. High quality, mechanically stable macroporous polymerics are now manufactured at much larger scales than the equivalent silica based reversed phase media, and are particularly popular in situations where the stationary phase requires cleaning in place. The polystyrene based media are stable to sanitization by treatment with concentrated sodium hydroxide solution, or with steam.

The invention of novel column hardware and complex stationary phases would be fruitless without the hard labours of dedicated chromatographers in the development of their art. The likes of Gregor Mann[1.8], Henri Colin[1.9], Geoff Cox[1.9,1.10] and Roger Nicoud[1.11] have been relentless in the arena of process modelling and optimization for preparative separations, to name a but few.

The recent surge in the popularity of preparative HPLC is probably a result of a more general urgency in the chemical industry. In pharmaceutical, biotechnology and agrochemical companies there is a market-driven force to bring products through faster that has allowed preparative HPLC to find its own niche. It is true that the final purification step for many drugs in the pharmaceutical and biotechnology industry already involves chromatography. However, in all of these industries there are many failures along the development pipeline and the number of man-hours, or to use a more modern term, full-time equivalents (FTEs) wasted chasing lost leads is costly. Preparative HPLC provides a tool to generate more compounds,

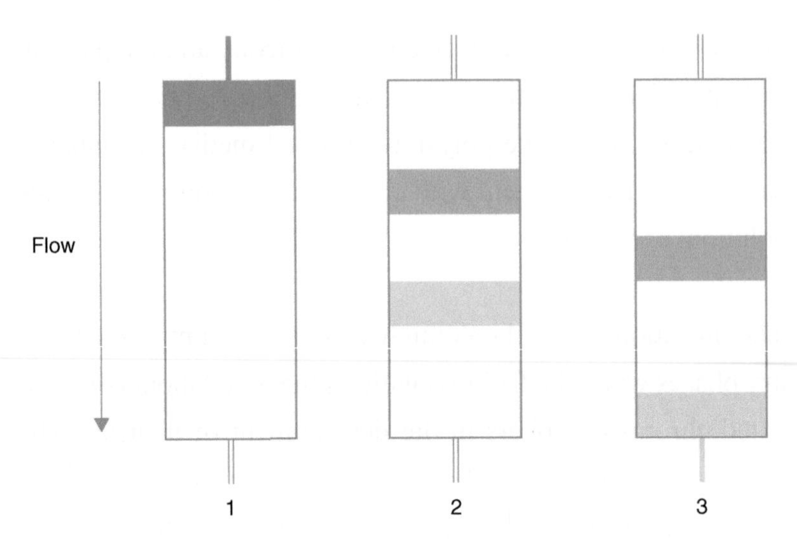

Figure 1.5

faster, from less pure products. It has been particularly valuable for chiral molecules where it can be difficult and time consuming to develop an asymmetric synthesis in comparison to a relatively simple separation by chiral HPLC. This mode of separation in particular has spawned the rapid development of SMB chromatography.

SMB is especially suited to the separation of binary mixtures, effectively splitting a chromatogram into two halves. The difficult step now is how to describe SMB in simplistic terms. Figure 1.5 helps to visualize the passage of a mixture of two components down a chromatography column. It would be convenient at position 2 to be able to remove each component separately whilst adding a constant feed to the top of the column.

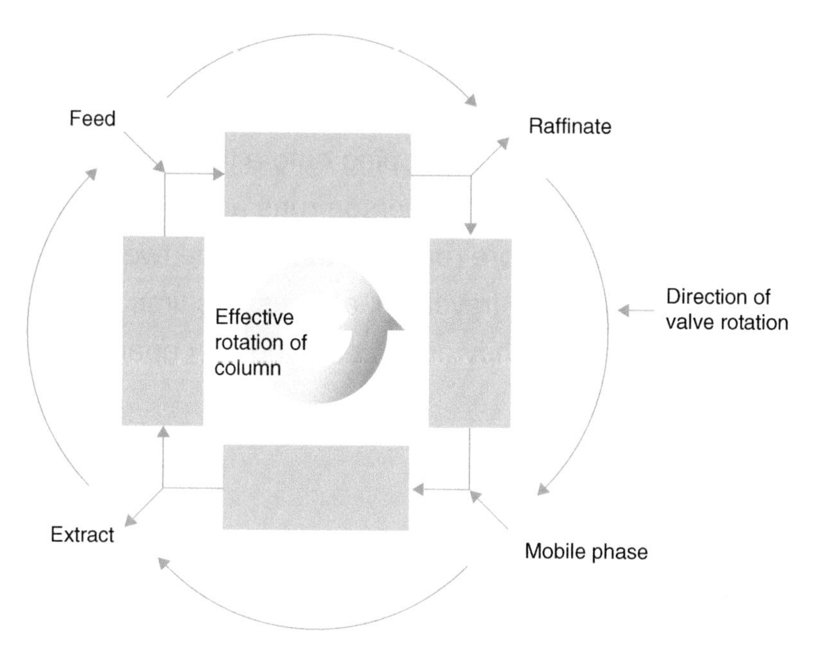

Figure 1.6

The optimum process for this binary separation would be to have fixed positions for the introduction of mobile phase and feed, and fixed collection points for the two components of the mixture whilst having the ability to move the stationary phase upwards. In practice it is impossible to engineer a system where the column bed moves, but it is possible to simulate the movement. Such a system is shown schematically in Figure 1.6 where four columns are set in sequence with four multi-port valves between the columns.

By the selective and carefully timed switching of the valves in a clockwise direction, the positions of feed, eluent, extract and

Confused?

Imagine two people stepping onto a travelator at the same time. One person runs and the other walks. Before reaching the end the two people leave the travelator, at the same time. Lo and behold, they are now a long way apart!

raffinate can be varied to allow the operator to simulate a bed moving anticlockwise – hence SMB.

The general technique is well-established and has been used for many years in the petrochemical[1.12] and sugar industries[1.13] in low pressure systems. The combination of SMB with preparative HPLC now allows the separation of mixtures with close running components[1.14]. The largest high pressure SMB system currently in operation, based at Lundbeck Pharmaceuticals in the UK, employs six HPLC columns of 80 cm diameter for the chiral purification Escitalopram.

Although SMB has been used predominantly for the separation of binary mixtures over recent years, it has also proved to be useful in the field of biotechnology[1.15]. One excellent example describes the purification of antibodies[1.16] and, more recently, the separation of nucleosides[1.17] has been discussed. The growing market for biopharmaceuticals will undoubtedly fuel a number of major developments in continuous chromatography in the years to come.

2

Fluid dynamics, mass transport and friction

In this chapter the mass and fluid transfer processes that dominate as a solvent passes over particles in a packed column bed are summarized in both physical and philosophical terms. To introduce some basic terminology and to put us on common ground, the liquid passing through column is referred to as the mobile phase whilst, in most cases, the solid particle is called the stationary phase.

Food for thought!

It will become apparent as an understanding of the philosophy is developed that in one particular mode of chromatography the stationary phase is not the solid particle.

Complex mathematical formulae will be minimized here for the purpose of simplicity since there are numerous texts that deal with detailed theory of mass transport in chromatography[2.1,2.2]. The flow of mobile phase through a packed column bed is shown schematically in Figure 2.1. There are two transport mechanisms in progress. Firstly, the convectional flow around the particles; and secondly, the diffusion in and out of the pores of the stationary phase.

In order to describe mass transport effects it is necessary to have an understanding of the measurements used to quantify

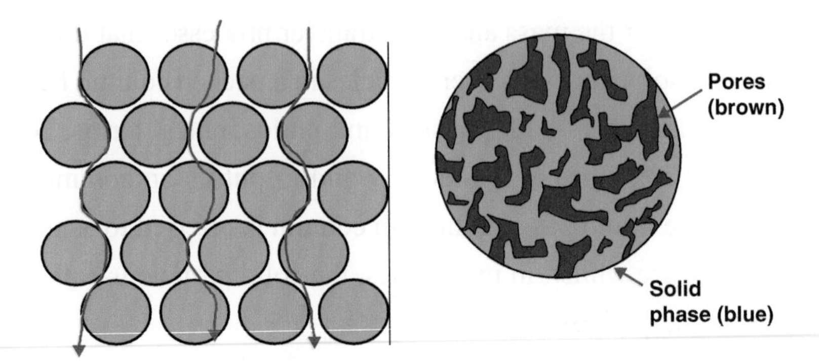

Figure 2.1

the efficiency of a chromatographic separation. Traditionally the term 'number of theoretical plates' is used to define the efficiency of the packed column bed. The mathematical derivations for plate theory were initially developed by Martin and Synge[2,3] and published in 1941.

Why plates?

This term actually originated in the petrochemicals industry and is derived from the oil refinery process, where an increased number of plates in a distillation column results in a more efficient separation.

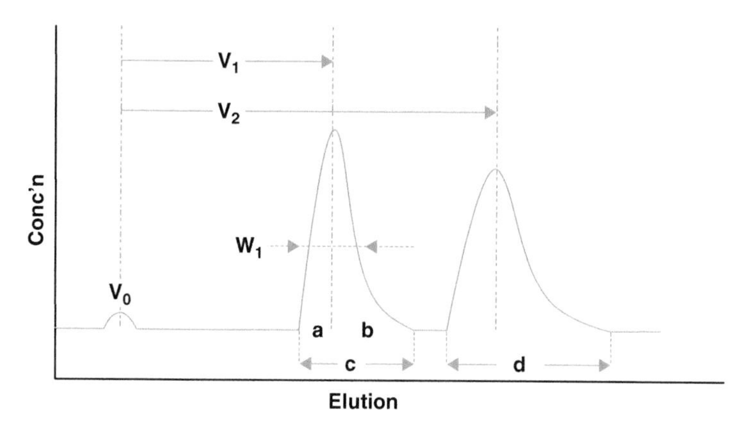

Figure 2.2

Figure 2.2 shows the measurements to be taken from a typical chromatogram containing two components.

As long as units are consistent throughout, the measurements can be recorded as time or volume. The measurement V_0 is the time taken for a non-retained component to travel from the injection port to the detector. The elution positions of the retained components, V_1 and V_2, respectively and the width of the first peak at half height, $W_{1/2}$ are used to calculate the number of theoretical plates, N and the separation factor, α.

$$N = 5.54\left(\frac{V_1}{W_{1/2}}\right)^2$$

The separation factor α, is effectively a measure of the degree of separation of two components. This term is derived from the capacity factor, k', of the corresponding peaks and is simplified to produce the following equation.

$$\alpha = \frac{V_2}{V_1}$$

The measurements a, b, c and d on the chromatogram can be used to calculate peak asymmetry, A and resolution, R.

$$A = \frac{b}{a}$$

$$R = \frac{2(V_2 - V_1)}{(c + d)}$$

Another term commonly employed is the height equivalent to a theoretical plate, HETP, which is simply the length of the column divided by the number of theoretical plates, N.

The path of a molecule dissolved in the solvent passing through a packed bed is fraught with obstacles. This individual entity will have to traverse the tortuous path around the stationary phase, where the number of potential routes is numerous and at some point during that journey it may have to seek out the most inaccessible site at the centre of that particle.

Picture this?

This can be likened to a mere mortal travelling through a dense forest with the task of having to climb the highest of trees. Imagine filling the forest with an entire army that have all been instilled with the same mission, add an enemy battalion or two then flood the forest with a fast flowing river!

In the presence of similar molecules and impurities, that molecule will also have to compete for the interactive sites on the surface of the stationary phase. The first scientist to assess the composite effects of mass transport in a chromatographic column from a chemical engineering perspective was JJ van Deemter[2.4] in the early 1950s. In doing so he derived a more dynamic equation for the HETP which, in simplified form, can be represented as:

$$\mathrm{HETP} = \mathrm{A} + \frac{\mathrm{B}}{\nu} + \mathrm{C}\nu$$

where ν is the mobile phase velocity and A, B and C are the influences controlling band broadening.

Did you know?

van Deemter was actually a physicist who applied a knowledge of packed beds in chemical processes to derive his classic equation. He received a memorial medal in 1978 honouring the 75th anniversary of the discovery of chromatography.

Component A of the equation encompasses the differing lengths of the tortuous paths taken by the solute molecules that ultimately leads to brand broadening. Band broadening caused by longitudinal diffusion is accounted for by component B, which in simple terms suggests that the less time a molecule spends in the column, the better. However, this component is counteracted by the resistance to mass transfer brought about by slow diffusion within the stationary phase and by the physical interaction with the surface of the media. As a consequence, high flow rates will also lead to band broadening and the resultant third component of the equation, C, is probably the most influential in flow rate optimization.

The optimization of flow rate is best represented graphically. Thus a plot of HETP versus flow rate will generate a graph similar to that shown in Figure 2.3.

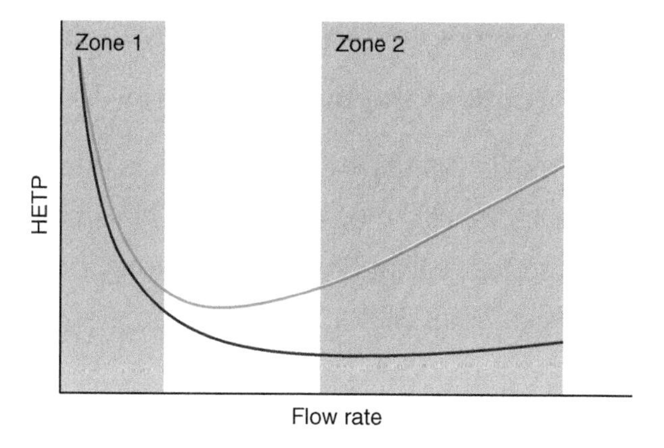

Figure 2.3

For purpose of illustration, the green line represents the results of a typical van Deemter plot. In Zone 1 the low flow rate allows extensive longitudinal diffusion, which ultimately will result in diffusion against the direction of flow. At high flow rates shown in Zone 2, the decreased efficiency is a result of comparatively slow mass transfer.

The blue line represents a situation where mass transport is relatively efficient. This might be observed when particle size is small, pore structure is large and the molecular dimensions of the analyte are small. In an analytical sense this looks good, but don't be fooled. Large pore size leads to low surface area, which consequently leads to comparatively lower loading capacity. For an efficient preparative separation where the objective is to maximize loading and minimize HETP it is always worth considering an investigation of pore size dimensions.

Think of the bigger picture!

Human Insulin contains 54 amino acids and has a relative molecular mass of 5808. This small globular protein will purify quite efficiently on stationary phases with 100 Å mean pore size.

Salmon Calcitonin has 32 amino acids and a relative molecular mass of 3432. This comparatively large peptide is in fact bigger than Insulin! Try >200 Å pore size – you might be surprised.

So! Get the stationary phase particle size as small as possible, get the pore size optimized and you're sailing. Slow down – it's not quite so simple. There are other major effects that have to be considered when scaling a preparative separation. The friction caused by the eluent passing over stationary phase particles generates heat, which in turn reduces the viscosity of the solvent. Cooling by conduction in the vicinity of the column walls reduces the viscosity of the solvent close to the wall in comparison to that at the centre of the column. Consequently, the solvent at the centre of the column is now travelling at a higher flow rate than that at the column walls, resulting in a parabolic flow profile, and subsequently, to band broadening.

In practice this column wall effect is particularly dominant in column diameters of 5 to 20 cm. This is explained schematically in Figure 2.4, where the dashed line represents the flow profile, or solvent front of the eluent and the grey area represents the depth of cooling by conduction. Cooling by conduction will penetrate radially to approximately the same depth with some variation due to the thickness of the column walls. For small diameter columns the effect is not observed since the whole diameter is cooled by conduction. As the column diameter increases the effect worsens. However, as the diameter gets much larger the effective depth of the cooling gets smaller in proportion to the diameter, so the effect on band broadening is lessened.

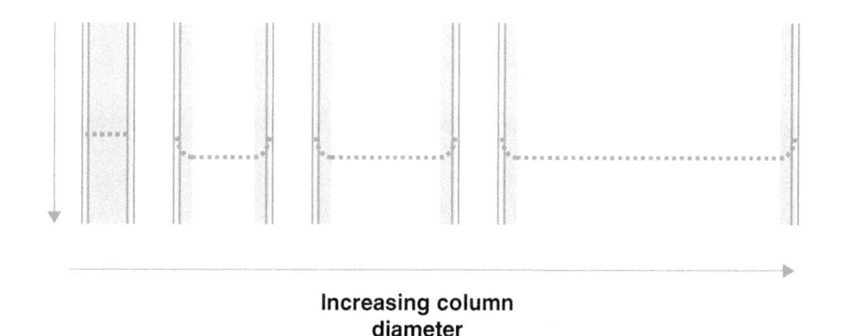

Figure 2.4

The effect can be counteracted by heating the column walls and nowadays it is commonplace to use jacketed columns. If a jacketed column is not available, the effect can be minimized by either lagging the column or by winding a tube around the column and circulating a heating fluid in the direction of the outlet to the inlet of the column.

3

Modes of chromatographic separation

Chapter Three

The most commonly utilized modes of chromatography employed preparatively at high pressure will be summarized in this chapter. The mode of chromatography is defined by the mechanism of interaction between the analyte and the stationary phase. Over one hundred years ago Mikhail Tswett[3.1] discovered that differing solvents selectively desorbed pigments from the leaves of plants. Contrary to the common belief that different pigments were soluble in different solvents, he correctly deduced that the selective desorption was due to variation in the interaction of the carotenoids with the cellulose based substrate of the leaves. To demonstrate his theory, Tswett logically chose filter paper as his stationary phase and this hydrophilic interaction between analyte and stationary phase is what we now refer to as 'Normal Phase Chromatography'.

Before proceeding with a discussion on the individual modes of separation it is worth reviewing a brief glossary of terms in association with the stationary phase properties. A simple strategy for selection of the stationary phase will naturally follow, in order to help the reader to choose a supplier wisely.

There are two main classes of polymeric support utilized in the preparation of stationary phases for all modes of chromatography, namely silica and polystyrene-divinylbenzene copolymer. For simplicity, from this point on, the former will be referred to as silica and the latter class as polymeric.

Particle size

The smaller the particle size the more efficient a separation will be. However, the smaller the particle the higher the back pressure in the packed column, which limits the practical size that can be used due to pressure limitations of the hardware. Typically, most equipment will be capable of withstanding the back pressures generated by 5 μm particles at optimized flow rates.

In theory, the more uniform spherical particle will form a better packed bed due to the additional ease of hexagonal close packing. In practice, the bed does pack better but there are more important underlying factors that control the efficiency of a separation including the surface morphology, pore structure and the pore size distribution. When considering spherical silica based particles there are few manufacturers that come close to supplying monodispersed stationary phases. In general the materials are prepared with a broad particle size distribution and subsequently classified to coincide with the manufacturer's specification. Conversely, polymeric stationary phases can be prepared in monodispersed form and this is one reason polymerics are theoretically cheaper when manufactured at large scale, though this may not always be reflected in the price.

Pore structure, size and surface area

The pore structure and size is critical for efficient chromatography and this is borne out by the fact that many supports of identical particle size, and nominally of similar pore size, often display different characteristics. As discussed in Chapter 2, van Deemter's theory suggests that the most critical attribute of a stationary phase is the ease of mass transfer through the pores of the particle.

The control of pore size and structure during preparation is a complex process that has been mastered by few manufacturers, but it is true to say that pore size distribution and uniformity is more readily achieved in silica particles than it is in polymerics. This is not to say that polymerics are inferior, because in contrast these stationary phases have a more uniform particle size.

Remember the forest?

The branches of an Oak are more diverse than those of a Spruce!

In practice, the less efficient mass transfer within polymeric stationary phases results in a lower optimum flow rate being obtained following a van Deemter plot. It is essential that the data required for a van Deemter plot is collected under DAC conditions similar to those anticipated for preparative operation. This is because the difference between silica and polymeric stationary phases is less pronounced when comparing pre-packed columns. The pre-packed hardware is packed at much higher pressures than DAC columns, which results in deformation of the polymeric particles. This deformation of the polymeric stationary phase reduces the interstitial spaces, giving the effect of having smaller particles.

When choosing a stationary phase, whether silica or polymer based, it is important firstly to choose the correct pore size to suit the analyte under investigation. Once the pore size is chosen it will soon become apparent that a range of stationary phases with varying surface areas is available. Clearly the larger the surface area the higher the capacity, but be aware that for a given pore size the higher the surface area the more fragile the stationary phase particle. As a result, you may have higher capacity but a shorter lifetime since the stationary phase will effectively crush under DAC.

At first glance the specifications of many stationary phases in a particular mode of separation may appear to be the same. For example, in reversed phase chromatography there are many C_{18} based media that will have the same carbon load, the same pore size distribution, and so on. However, stationary phases from different manufacturers will behave differently, so it is important to reiterate the need to screen seemingly identical materials to get the optimum separation required.

3.1 Normal Phase Chromatography

The IUPAC Compendium of Chemical Technology defines 'Normal Phase' as 'an elution procedure in which the stationary phase is more polar than the mobile phase'. In practice, the most widely used stationary phases for preparative HPLC are based on silica and the polarity of the underlying silyl ether and silanol provides the required hydrophilic surface. Amino and cyano bonded silica are also commonly used in normal phase mode though the latter also has some reversed phase properties. The predominant mechanism of interaction is hydrogen bonding. However, the silanol is mildly acidic so the silica surface will also have mild cation exchange properties.

Normal phase chromatography is commonly used to purify small organic molecules so the technique is regularly employed in the pharmaceutical and agrochemical industries. The progression from flash chromatography to preparative HPLC has been easy to comprehend for traditional organic chemists and this has probably driven the adoption of the technique for many pilot and plant scale separations. A concomitant transition from irregular particles to mechanically robust spherical stationary phase particles has also been fuelled by the increased demand. As stated earlier, irregular silica based particles are relatively fragile and, as such, are prone to degradation and generation of fines.

A sophisticated and complex Ball Mill!

I have known the performance of irregular silica particles to improve with time and it has been postulated that this is due to erosion of the media to such an extent that the stationary phase has eventually become spherical.

Figure 1.4 in Chapter 1 shows a scanning electron micrograph (SEM) of typical irregular and spherical silica particles for comparison.

Normal phase chromatography is typically, though not exclusively, performed in organic solvents. Retaining solvents are generally hydrocarbons, such as hexane, heptane or cyclohexane; or halocarbons including dichloromethane and chloroform. The eluting mobile phase is generally a miscible hydrophilic solvent, such as alcohols and esters. Organic acids are often added to minimize any cation exchange effects arising from acidic silanol groups on the silica surface.

As stated above, the utility of silica based stationary phases does not limit its use to organic mobile phases. For many years it has been commonplace in flash chromatography to use aqueous solvents to elute analytes from silica based media. Isocratic elution with mixtures of butanol, acetic acid and water is standard protocol for the separation of amino acids and a carefully prepared combination of methanol, chloroform and water is useful for general organic compounds. Peptides are also readily purified by gradient elution on normal phase silica, moving from acetonitrile to aqueous mobile phase[3.2]. This technique is particularly useful for extremely hydrophilic peptides that are not strongly retained on reversed phase media.

It is also useful to note that the hydrophilic 'alcohol coated' stationary phases originally designed for size exclusion chromatography such as Zorbax GF250 and PL-aquagel also work well as normal phase media. The advantage of these lies in the increased stability of the stationary phases at high pH.

3.2 Reversed Phase Chromatography

The IUPAC definition understandably states that reversed phase chromatography is 'an elution procedure in which the mobile phase is significantly more polar than the stationary phase'. This is a somewhat simplistic statement that covers a wealth of possibilities mirrored by the broad range of reversed stationary phases available, to what is without doubt the most commonly applied mode of preparative HPLC. A brief survey suggests that there are over five hundred different reversed phase materials commercially available. However, the eventual scale of operation may limit the number of suitable stationary phases to less than ten.

Reversed phase media are designed to operate in aqueous buffers where the hydrophobic adsorption between analyte and stationary phase is disturbed by an increasing concentration of a miscible organic solvent. The most common application of reversed phase HPLC is in the purification of peptides and proteins where the analyte is often desorbed under gradient elution. The most common aqueous buffer is water containing low levels of trifluoroacetic acid (typically 0.1% v/v) and the eluting mobile phase – or polar modifier as it is often termed – is usually acetonitrile containing trifluoroacetic acid. The trifluoroacetic acid serves two purposes: firstly, it forms a strong ion-pair with the analyte thus counteracting the cation exchange with any free silanol groups on silica supports; and secondly, by virtue of the ion-pair, it reduces the conformational variations in peptides and proteins consequently improving peak shape. The added advantage of trifluoroacetic acid lies in its transparency to ultraviolet absorption, the major technique used for the detection of peptides, proteins and many other molecules. Other acidic buffers including phosphate, acetate and hydrochloride are also commonplace. Ethanol, methanol, tetrahydrofuran, propionitrile and propan-2-ol are also regularly used as polar modifiers but these do not normally have the broad resolving power of acetonitrile.

The original reversed phase media were based upon silica reacted with chloroalkylsilanes to covalently attach hydrophobic ligands such as dimethyloctadecylsilane. The stationary phases prepared in the early stages of development of the technique suffered due to incomplete reaction of the silica surface. Residual silanol groups on the surface (Figure 3.1) are both hydrophilic and mildly acidic, resulting in a mixed interaction with the analyte and consequently leading to band broadening. This was particularly prominent for basic compounds due to cation exchange with the free silanol groups. Understandably, the first major development in this mode of separation was to block the unreacted groups with smaller chloroalkylsilanes such as chlorodimethylethylsilane in a process commonly termed 'end-capping'.

There have been many developments over the years, including improvements in the 'wettability' of the surface by introducing polar groups between the alkyl chain and the silica. There is also a range of alkyl chain lengths available but the most popular remain to be C_8 and C_{18}.

Silica based stationary phases are unstable to alkaline conditions due to loss of the alkyl chain and dissolution of the silica

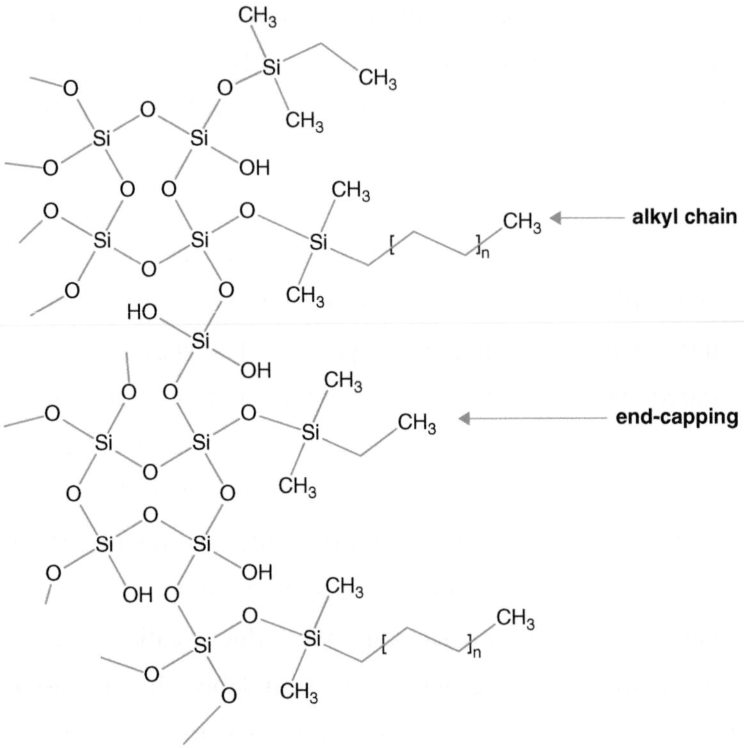

Figure 3.1

so long-term operation at high pH can be limited. The introduction of macroporous polymeric stationary phases has broadened the pH range for reversed phase HPLC since these media are stabile from pH 1 to pH 14. The most commonly used polymeric reversed stationary phases are based on copolymers of styrene and divinyl benzene. The hydrophobicity of these media is derived from the polymer backbone and pendant benzene rings, so the elution profiles are often expected to be different from a typical C_{18} coated silica for instance. In reality, for most analytes the elution profiles, elution order and elution times observed on high quality polymerics is often similar to the results obtained from high quality silica based reversed stationary phases.

Analysis and subsequent purification of peptides at high pH can be a useful tool for the chromatographer. It is not uncommon for impurities co-eluting at low pH to be completely resolved at high pH and the ability to operate at high pH is particularly advantageous for the separation of peptides rich in aspartic and glutamic acid. Probably one of the major advantages of polymerics lies in the extreme base stability, which allows cleaning in place (CIP) using concentrated sodium hydroxide solutions. Though many manufacturers of silica based reversed phase media claim to have enhanced stability in alkaline conditions, none of these will actually stand up to prolonged and regular CIP.

3.3 Chiral Separations

The importance of this mode of separation lies in the shear scale of the market for optically pure molecules. The sales of single enantiomer chiral drugs is currently U$180 billion per annum and ~65% of active pharmaceutical ingredients (APIs) currently in development have at least one chiral centre. Apart from the pharmaceutical industry there is also considerable potential in the chiral agrochemicals market.

Chiral purity is generally introduced in one of three ways:

- chiral purification
- asymmetric chemistry
- biotransformation.

In order to fully explain chiral chromatography it is necessary to have an understanding of chirality itself. It is worth describing how chiral separations work in order to help to dispel some of the myths surrounding the complexity of this mode of separation. A few brief definitions and formulae required to understand the summary are listed below.

Chirality

A *chiral* object cannot be superimposed on its mirror image whilst an *achiral* object can be superimposed on its mirror image. The two forms of a chiral molecule are known as *enantiomers*. A single chiral isomer is said to be enantiomerically or optically pure. Whereas an equimolar mixture of enantiomers is said to be *racemic*.

The central point of chirality in a molecule is known as the *stereo-centre* and *diastereoisomers* occur when there is more than one *stereo-centre* in a molecule. *Stereo-centres* are given the absolute configurations of *Rectus* (R) and *Sinister* (S). Diastereoisomers that differ in *absolute configuration* at the *stereo-centres* are called *epimers*. Diastereoisomers differ in conformation so it is actually possible to purify these mixtures by normal phase and reversed phase HPLC. However, better separation factors are often obtained using chiral stationary phases.

Optical purity

Optical purity is defined by the terms *enantiomeric excess*(ee) and *diastereoisomeric excess*(de) given by the formula shown below.

$$\%ee = \frac{m_a - m_b}{m_a + m_b} \times 100$$

This term is related to the two chiral components of a mixture and as such does not describe the total purity by area. This is demonstrated diagrammatically in Figure 3.2.

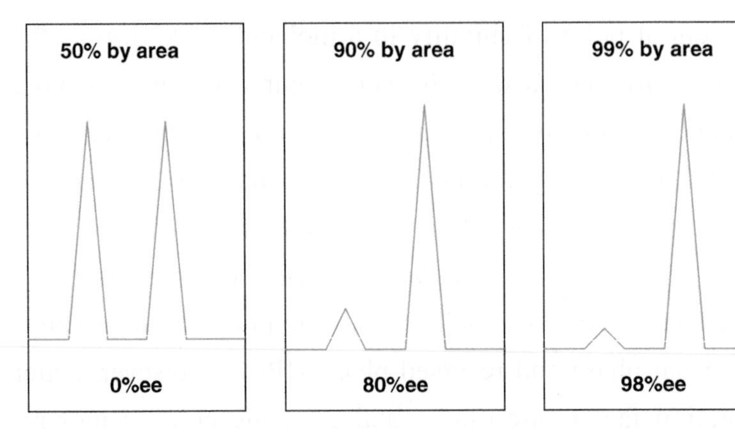

Figure 3.2

Separation of enantiomers

Enantiomers are traditionally separated by crystallization of diastereoisomeric salts or by diastereoisomeric derivatization followed by crystallization. This technique is somewhat limited if there is no acidic or basic nature or no point of derivatization. In addition, it can be a laborious process to optimize the crystallization process.

More than 30 years ago, Bill Pirkle, the recognized inventor of modern chiral HPLC, realized that it may be possible to effect a chromatographic separation of enantiomers by use of chiral selectors (or ligands) bound to a silica matrix[3.3,3.4]. There has been a phenomenal amount of development in chiral stationary phases over subsequent years but, a relatively small number of

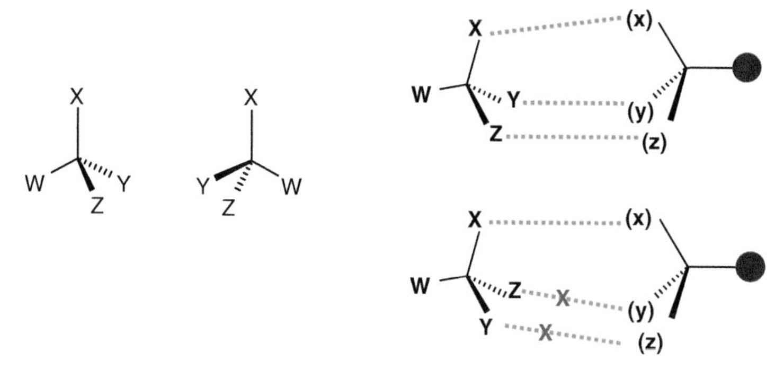

Figure 3.3

these has led to domination of the preparative market by two or three companies over the last 20 years.

The interaction between the analyte and the stationary phase in chiral chromatography can involve characteristics of molecular size, hydrophilic, hydrophobic and ion-exchange mechanisms, often simultaneously. In general, the surface of the stationary phase is covered with a chiral ligand that will bind preferentially with one of the enantiomers of the mixture. The mechanism of preferential binding, defined by Dalgliesh[3.4] in 1952 as the 'three point rule', is shown schematically in Figure 3.3. The left-hand side of the diagram shows a tetrahedral object and its non-super-imposable mirror image, or optical isomer, based upon WXYZ. The right hand side of the diagram represents an immobilized structure based upon (x), (y) and (z). Imagining the groups WXYZ to be the functional moieties of an enantiomeric mixture and (x)(y)(z) to be the interacting groups of an immobilized chiral ligand, it is easy to observe that the binding of one enantiomer will be preferred over the other.

This 'three point rule' was further ratified by Pirkle and House[3.5] who stated that 'there must be at least three simultaneous interactions between a chiral stationary phase and a solute enantiomer and one of these must be stereochemically dependent, if chiral resolution is to be effected'. Some of the earliest chiral stationary phases developed by Pirkle contain various binding sites including, π-electron donors, π-electron acceptors and amides or esters for hydrogen bonding. Many other chiral ligands have been utilized over the ensuing years including proteins, cyclodextrins, antibiotics and crown ethers.

Although they are extremely useful analytically, the protein based stationary phases[3.6,3.7] have found little application in preparative HPLC because they suffer from low loading capacity, due primarily to the low number of active sites. The natural macrocylic molecules Cyclodextrin[3.8,3.9] and antibiotics such as Vancomycin[3.10] have shown some promise. Synthetic chiral crown ethers[3.11] are particularly useful for the separation of chiral primary amines.

More recently, polymeric tartaric derivatives[3.12] covalently bound to silica are proving to be useful in preparative applications due to enhanced physical and chemical stability. However, the most extensively used media by far are based upon modified amylase and cellulose polymers adsorbed to the surface of silica[3.13]. When describing the basic concept of the chiral stationary phase the word 'covered' was used deliberately since in this case, until recently, the ligand was not covalently bound to the surface of the stationary phase. The second example of process development described in Chapter 5 involves a chiral separation on this type of chiral stationary phase.

There have been several useful reviews written on chiral chromatography over recent years but one of the most comprehensive was written by Levin and Abu-Lafi in 1993[3.14].

3.4 Ion Exchange Chromatography

The majority of applications utilizing ion exchange as a mode of chromatography are generally carried out at low to medium pressure. This approach is commonplace in the biotechnology arena where it is predominantly used for the purification of biological macromolecules such as proteins. However, it is worth summarising the high pressure application of the technique since it is particularly useful for the purification of synthetic oligonucleotides. The DNA therapeutics market is extremely buoyant at present and most pharmaceutical companies have their foot in the door or are at least collaborating with a group or company investigating oligonucleotide based drugs. The predicted market potential ranges from hundreds of kilograms to multi-tonne requirements. Therefore the subsequent growth in synthesis will be mirrored by a concurrent growth in purification developments.

Ion exchange chromatography, as the name suggests, separates molecules by taking advantage of a charge differential. There are understandably two approaches to ion exchange separations, so both cation and anion exchange stationary phases are commercially available. The stationary phases are predominantly polymer based and functionalized with acidic groups to produce cation exchange materials, or basic groups to produce anion exchangers. Cation exchange media are generally of carboxylic acid or sulfonic acid functionality to provide weak and strong cation exchangers, respectively. Conversely, the weak and strong anion exchange media are typically functionalized as tertiary or quaternary amines. Ion exchange chromatography is generally carried out in aqueous environment where the charged components of a mixture are desorbed from the stationary phase by either changing the pH of the eluent or by adding a stronger counter-ion and effectively displacing the analyte.

The technique for anion exchange based separation is described diagrammatically in Figure 3.4 as an example. The diagram shows the separation of a mixture containing species carrying different levels of negative charge. Initially both species are bound under none eluting conditions and are then selectively

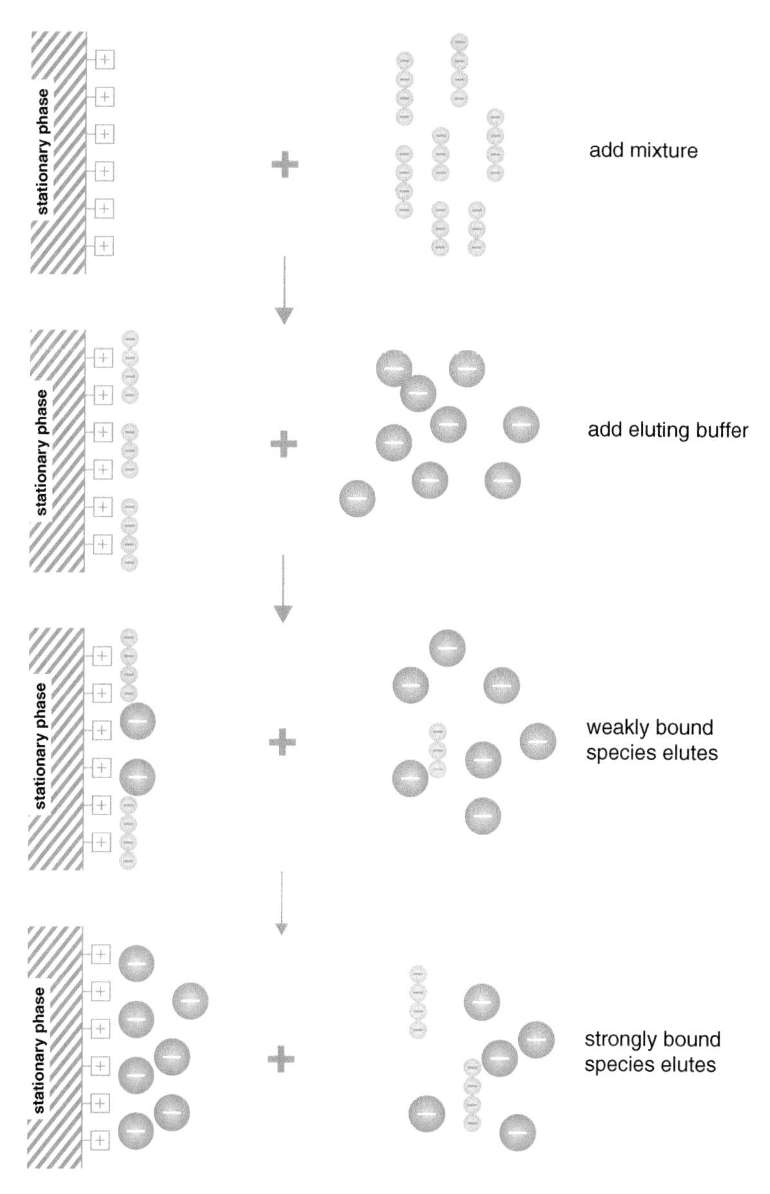
stationary phase
add mixture
stationary phase
add eluting buffer
stationary phase
weakly bound
species elutes
stationary phase
strongly bound
species elutes

Figure 3.4

desorbed as the concentration of a more strongly bound anion is increased.

The mixture could be eluted by decreasing the pH of the eluent. However, it is usually easier to elute by increased salt concentration since this is simpler to control.

Invest in a conductivity meter!

If the ionic strength of the sample solution is much higher than that of the starting conditions the mixture will not bind. A simple dilution can save a lot of lost time and reduce frustration.

The basic structure of an oligonucleotide is shown in Figure 3.5. The major impurities in synthetic oligonucleotides are usually nucleoside deletions and occasionally insertions. The impurities will therefore have a different charge density compared to the target molecule. The longer, more highly charged molecules have a stronger retention, so deletions will elute earlier than the target molecule which will, in turn, elute earlier than nucleoside insertions.

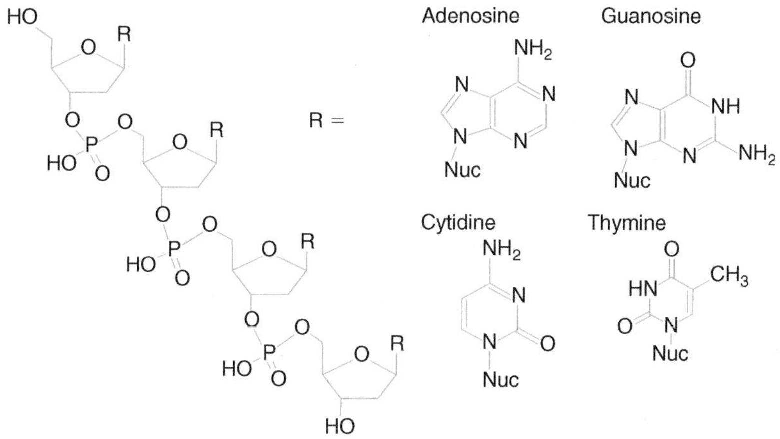

Figure 3.5

Oligonucleotides are normally purified by anion exchange chromatography. The crude mixture is immobilized at high pH and often eluted by increased concentration of a salt, such as sodium chloride. It is advantageous to add 15–20% of a miscible organic solvent, such as acetonitrile or ethanol, to counteract any reversed phase interaction with the underlying polymer backbone. Figure 3.6 shows a typical example of a crude synthetic oligonucleotide containing 18 bases, separated on a strong anion exchange resin.

When purifying oligonucleotides it is particularly useful to use sample self displacement chromatography since the required component of the mixture is generally the later eluting moiety. With this approach the column loading is increased to such a point that the more strongly retained component displaces the

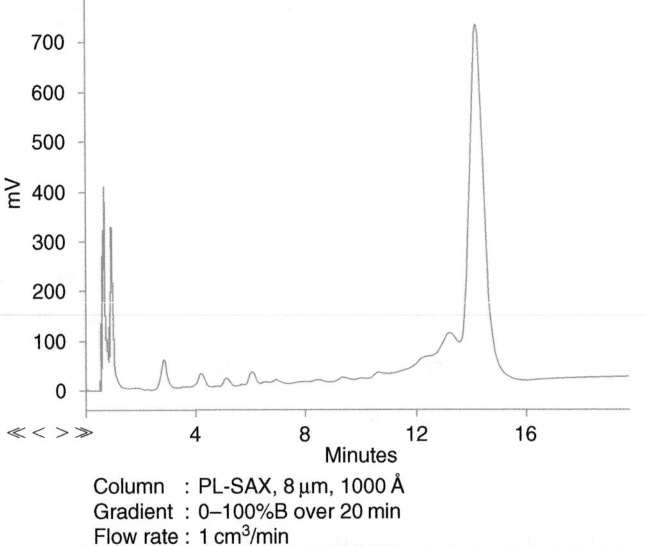

Figure 3.6

less strongly adsorbed species. The technique is described in detail in Chapter 5.

One of the drawbacks of ion exchange chromatography is the need for a secondary technique to remove inorganic salts from the purified product. Desalting can often be performed by ultrafiltration, solid phase extraction or by gel filtration. The latter mode of separation is described briefly in Section 3.5.

3.5 Exclusion Chromatography

This technique is often referred to under several headings such as gel filtration, gel permeation, or size exclusion chromatography. In its simplest form, gel filtration, it is often used to desalt solutions and is regularly used to compliment ion exchange separations. In more complex separations it can be used to separate proteins of different size or shape. This mode of separation is rarely performed at high pressure and is often reserved for separating mixtures, as the title suggests, by taking advantage of the differing size.

Why discuss this approach if it is limited to low pressure operation? The ongoing and future development of recombinant processes for the preparation of biopharmaceuticals will require a concurrent high performance approach to separation of proteins from fermentation broths and from salts. The productivity of current approaches using low pressure systems or ultrafiltration will be insufficient to cope with the long-term demand. It is inevitable that manufacturers will develop high performance techniques and high pressure packing media. It is possible to envisage the use of continuous techniques such as SMB chromatography in size exclusion mode[3.15].

The column packing media used in size exclusion chromatography is available in various pore sizes, designed to exclude larger molecules. For simple desalting of a protein the separation works by excluding the biological macromolecule from the pores of the packing media and allowing the salt to penetrate the pores, delaying its elution. Clearly the protein is not retained and simply follows the path of convectional flow through the interstitial spaces between the particles. In a slightly more complex situation, often referred to as gel permeation chromatography, the particles have varying pore sizes designed to allow penetration by the large species. This technique can be used to separate proteins and protein aggregates of various sizes and, as such, is likely to be important in the growing recombinant market.

Packing Media! What happened to 'Stationary Phase'?

In size exclusion chromatography the 'Stationary Phase' is the solvent inside the pores of the column packing media. Conversely the 'Mobile Phase' is the solvent in the interstitial spaces.

3.6 Affinity Chromatography

Again, until recently, affinity chromatography has been limited to low pressure operations. However, as described above for size exclusion, affinity chromatography has a special place in the purification of biological macromolecules. The growing popularity of biopharmaceuticals, especially those derived from recombinant processes, will almost certainly require major developments in high pressure stationary phases for affinity chromatography.

This mode of separation, as the name suggests, uses stationary phases with a special affinity for a specific analyte. The affinity ligand immobilized on the stationary phase varies dramatically from peptide, to protein, to oligonucleotide, to monoclonal antibody. In some cases the target molecule is labelled with an affinity tag to simplify the separation. This approach is common in the synthesis of recombinant proteins where the system can be engineered so that the target biomolecule expresses a tag such as polyhistidine. A stationary phase functionalized with aminodiacetic acid and nickel chelate is then used to fish out the required molecule by chelating with the polyhistidine tag.

Existing stationary phases used in this area are usually soft gels that often suffer from low loading capacity brought about by the inability of biological macromolecules to penetrate the matrix. The most likely progress in this arena over the forthcoming

Have you realized?

Chiral chromatography is a variant of Affinity Chromatography.

years will be in the development of macroporous media with high loading capacities, and the mechanical rigidity essential for operation at high pressure.

The technique of affinity chromatography was described in some detail by Lowe and Dean in a text published in 1974[3.16].

4

How to get started

Chapter Four

4.1 Packing a Column

For process scale operations it is normal for users to pack stationary phases in their own column hardware. The most common equipment available is based on the dynamic axial compression system invented by Couillard[4.1], subsequently assigned to, and first marketed by Prochrom (now NovaSep). The original patents on this technology have now expired so this type of column format is now available from a range of suppliers. In general, this simple but very effective technology involves the use of a moving piston as one of the column end fittings thereby allowing a constant compression of the packed bed with a dynamic removal of column voids. Examples of preparative DAC columns and HPLC systems are found in Figures 4.1 and 4.2. Figure 4.1 shows a 20 cm diameter DAC column, which can be seen on the left of the photograph. Figure 4.2 shows an industrial scale column with an internal diameter of 60 cm.

Other column designs have been developed, including radial compression[4.2] and annular expansion[4.3]. The radial compression system originally developed by Waters, and now marketed by Biotage, uses flexible polymeric cartridges packed with stationary phase. The cartridge is housed in a cylinder and the axial compression is generated by applying external pressure to the cartridge. In the annular expansion columns a tapered

Figure 4.1 Courtesy of Jerome Theobald, NovaSep SAS

Figure 4.2 Courtesy of Jerome Theobald, NovaSep SAS

rod runs through the column and compresses the stationary phase against the walls as the rod is pushed into the column, providing both axial and radial compression.

This text will concentrate on the methodology used to pack DAC based columns. There are two basic ways of packing these columns that are determined by the design of the column format. The two systems normally available use either a short column with a detachable column extension (Figure 4.3), the

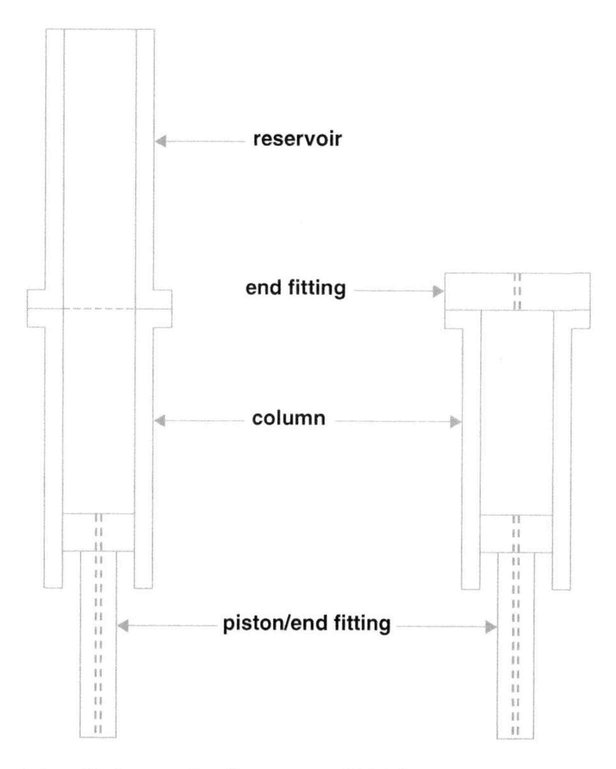

Figure 4.3 Schematic diagram of DAC column

combined volume of which is sufficient to hold the total slurry volume, or a long column, large enough to hold the total volume of the stationary phase slurry (Figure 4.4).

When using the column type shown in Figure 4.3 the reservoir, or column extension is left in place while the slurry is added and removed later. With the column type shown in Figure 4.4

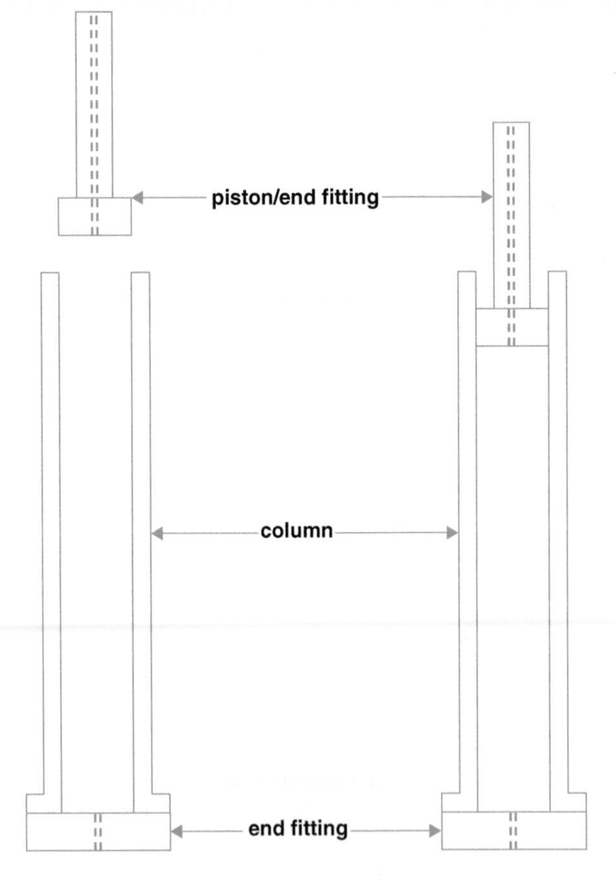

Figure 4.4 Schematic diagram of DAC column

the fixed end fitting is normally, though not exclusively, at the bottom of the column and the slurry is added prior to inserting the piston. In instances where the piston is at the bottom of the column the actual compression of the system moves and disturbs the bed often leading to less efficient packing.

Column packing protocol for column type 1 (Figure 4.3)

This protocol will concentrate on the methodology used to pack the column type shown in Figure 4.3. The basic layout normally includes a short column with a detachable column extension, the combined volume of which is sufficient to hold the total slurry volume.

When using the column type shown in Figure 4.3 the reservoir, or column extension is left in place while the slurry is added and reservoir removed later.

Slurry concentration

The recommended solvents for various stationary phases vary from one manufacturer to the next and will depend upon the stationary phase being packed. However, the concentration of stationary phase in solvent is generally in the region of 1 kg of stationary phase in 2–3 dm^3 of solvent for silica based stationary phases. Polymeric stationary phases are less dense than silica based media so a dilution of 1 kg in 4–6 dm^3 of solvent is recommended.

Slurry and column preparation

Calculate the total volume of the column plus the packing reservoir in order to determine the total volume (volume A) of the required slurry. Determine the amount of stationary phase required to pack the required bed length (the packing density of polymerics is generally ~0.3 g/cm^3 and the density of silica based media is generally ~0.6 g/cm^3). Add the appropriate amount of stationary phase to a suitable container and dilute with the chosen solvent to 'volume A'. Ensure that the slurry is efficiently mixed either manually or by gentle mechanical agitation. Once a homogeneous slurry is obtained, halt the agitation and leave the slurry to stand overnight to allow the stationary phase to de-gas.

Set up the column with the packing reservoir in place. Attach a vacuum line and solvent collection vessel to the bottom of the column. Do not apply a vacuum at this stage. Re-suspend the stationary phase either manually or by gentle mechanical agitation then pour the slurry, without stopping, into the column-reservoir assembly. It is essential that the slurry is added as quickly as possible to avoid settling. A pause of even a few seconds during pouring the slurry will lead to a break in the column bed and subsequently to poor chromatography.

Apply a vacuum to the outlet of the column and draw off the solvent until it is a few millimetres above the bed surface. Remove the vacuum line and plug the bottom column outlet.

Remove the column reservoir and attach the column end fitting. Attach a waste solvent line to the top end fitting (with the bottom fitting still plugged). Compress the column until the packing pressure recommended by the column manufacturer is achieved (typically 60–100 bar). The column is now ready for equilibration in the solvents required for the separation. Although the position of the piston can often be locked it is actually best kept under constant pressure in order to maintain a well packed bed.

Column packing protocol for column type 2 (Figure 4.4)

Although the slurry concentration and process for the preparation of the slurry remains the same, the protocols for column packing differ slightly due to the general layout. In this system the column length accommodates the total slurry volume.

The bottom fitting of the column is plugged and the slurry poured in as before. The piston is then inserted with the top fitting attached to a waste solvent tank. As the piston pressure is increased the solvent in the column is excluded. It is not essential to plug the bottom fitting with either column design as long as the waste solvent, which is under pressure, is captured. However, plugging the bottom frit improves packing near to the top of the column bed as solvent is excluded.

In both cases the column should not be pressurized beyond the manufacturer's recommendations.

4.2 What and Where!

What use would a 'practical handbook' be if it did not tell you what you need and where to put it? The guidelines given here are provided purely as an indicative potential of the capacity of a certain column size and the estimated footprint the column and its ancillary equipment will occupy. The capacity of the stationary phase will depend upon the mode of chromatography and the method of operation. For instance, the loading capacity of reversed phase and chiral stationary phases is generally at least half that of a normal phase media. This is not meant to imply that normal phase chromatography is more

Table 4.1

Scale (kg/annum)	Column diameter (cm)
500	10
1125	15
2000	20
4500	30
10125	45
18000	60
50000	100

efficient because the overall efficiency should be based on productivity, raw material costs and waste disposal.

Table 4.1 summarizes the potential quantity of a crude mixture that could be processed annually in comparison to column diameter. This assumes a typical column loading of approximately 60 g per kg of stationary phase, which would not be unusual, and 24 h operation. It is important to note that the figures given are for throughput of crude starting material and that yield will depend upon the initial product purity.

As a guideline for the space required for preparative HPLC equipment the estimated footprint of the column, ancillaries and pumping system for various column diameters is shown in

Table 4.2

Column diameter (cm)	Foot print: Column	Foot print: HPLC	Location
10	0.25 m^2	0.6 × 1.5 m	Lab fume cupboard
15	0.64 m^2	0.6 × 1.5 m	Lab fume cupboard
20	1.44 m^2	1.5 × 2 m	Pilot plant
30	4 m^2	2 × 3 m	Plant
45	6.25 m^2	2 × 3 m	Plant
60	9 m^2	3 × 3 m	Plant
100	16 m^2	3 × 3 m	Plant

Table 4.2. This does not take account of the solvent supply system or fraction collection arrangements, which will be dependent upon the mode of separation.

The basic components of a preparative HPLC system shown in Figure 4.5 simplifies the overall process. In a more realistic form the colour coded schematic diagram of Figure 4.6 shows a typical plant layout for a facility housing a 30 cm diameter column. The section outlined in green covers solvent delivery, red is used for the post column solvent flow, sample feed is shown in blue and the stationary phase preparation area is in turquoise.

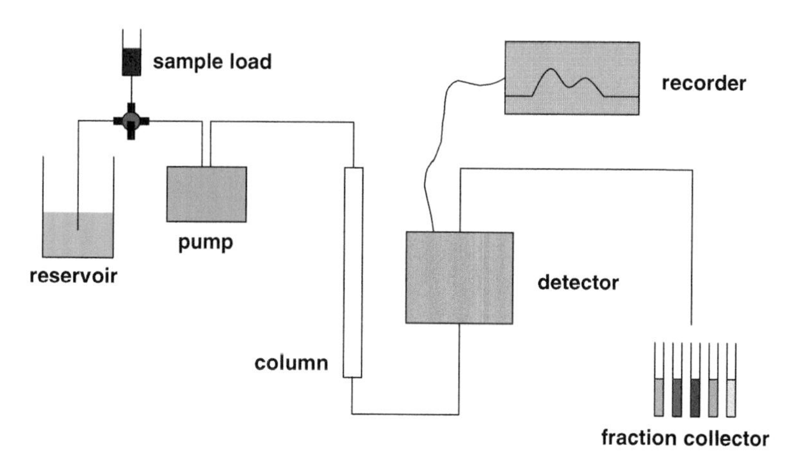

Figure 4.5

The system shown is capable of performing a quaternary gradient and shows all of the elements to be considered for industrial scale separations. For instance, a solvent reservoir size of 1–2 m^3 would allow continuous operation for several hours. Separations carried out at this scale will be extremely well-defined with respect to fraction size and point of collection so very few fraction collection vessels would be required.

4.3 Product Recovery

It would seem a rather pointless exercise planning the installation of a preparative HPLC system without considering the process to recover the product from the chromatographic eluent. There is no simple answer to the question since it will

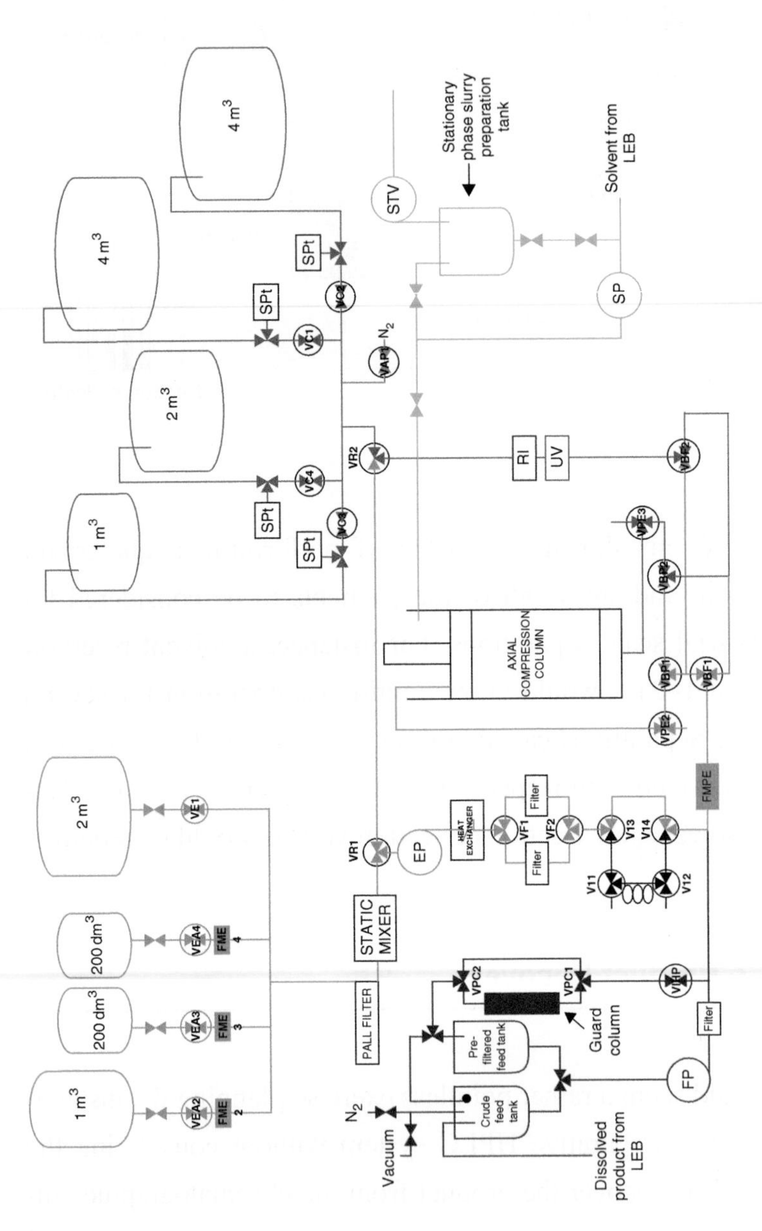
4 m3
4 m3
2 m3
1 m3
SPt
VC1
VC4
VR2
VAP1
N2
STV
Stationary phase slurry preparation tank
Solvent from LEB
SP
RI
UV
VBF2
VPE3
VBP2
AXIAL COMPRESSION COLUMN
VBP1
VBF1
VPE2
FMPE
2 m3
200 dm3
200 dm3
1 m3
VE1
VEA4
VEA3
VEA2
FME
VR1
EP
HEAT EXCHANGER
VF1
VF2
Filter
V13
V14
V11
V12
STATIC MIXER
PALL FILTER
VPC2
VPC1
VHP
Guard column
Pre-filtered feed tank
Crude feed tank
FP
Filter
N2
Vacuum
Dissolved product from LEB

Figure 4.6

depend on the nature of the product and the solvents to be removed.

For biological macromolecules such as peptides, proteins and oligonucleotides, which are generally purified by reversed phase chromatography, the difficulty lies in the removal of water. Lyophilization is a very mild option on a small to intermediate scale, but this becomes impractical and expensive when the volumes reach several hundred litres. The validation of a freeze drying process can also be very complex and the process reproducibility is poor, often leading to varying levels of residual solvent in the final product. For cGMP processes lyophilization from bottles and flasks does not provide sufficient reproducibility and instruments with temperature controlled shelves can be expensive. Wiped film evaporation is a useful complimentary technique for concentration of solutions prior to lyophilization.

Ready packaged!

When lyophilising from trays in a shelf freeze dryer try transferring the solution to Gortex bags before lyophilization.

Spray drying can be a useful option for large volumes of aqueous solvents and is used regularly for the recovery of peptides and proteins. However, before investing in spray drying equipment it is worth confirming the stability of the product to the process. When using reversed phase chromatography for purification of small organics it is often easy to remove the polar modifier by rotary evaporation and extract the product using an immiscible solvent. The predominantly organic solvents used for normal phase separations can be removed by traditional means of evaporation including wiped film and rotary evaporators.

4.4 Productivity

For those considering the development of a preparative separation with the long-term objective of using preparative HPLC as part of a manufacturing route, the costs associated with the routine operation will be a major focus of the decision-making process. The ultimate value of the product and the costs of developing alternative purification methods will have to be taken into account and for many targets there may be no alternative to HPLC. For the purification of synthetic peptides there is no single technique that will provide the target molecule at a higher purity than reversed phase chromatography. The same can be said when using a combination of anion cxchangc and reversed phase techniques for the purification of synthetic oligonucleotides. At the other end of the scale, separation of a racemic mixture by chiral chromatography can sometimes be avoided by an asymmetric synthesis. However, if the time and person-power taken to develop asymmetric chemistry is taken into account, it may be worth reconsidering preparative HPLC since the establishment of a chromatographic process is generally extremely rapid by comparison.

The productivity of a typical reversed phase purification of a peptide is summarized in Table 4.3. For this particular separation the column loading is relatively low and a large number of separations are carried out in order to deliver the annual production target. On the face of it, the productivity of this separation is poor; in practice the column size was chosen to balance the output with the rate of synthesis of crude peptide. The actual combined costs of synthesis and purification for this particular peptide was only a tiny fraction of the value of the formulated drug. On occasions it may be better to purify a valuable product in smaller portions rather than risk a proportionally high loss from failed separation.

Table 4.4 shows the comparative data for purification of a traditional small molecule by normal phase chromatography.

Table 4.3

Column diameter	15 cm
Column length	25 cm
Stationary phase	3.5 kg
Eluent (water/acetonitrile)	78:22
Column loading	40 g
Cycle time	60 min
Solvent consumption/kg of crude processed	600 dm^3
Annual production	40 kg

Table 4.4

Column diameter	100 cm
Column length	25 cm
Stationary phase	155 kg
Eluent (organic)	100%
Column loading	15.5 kg
Cycle time	60 min
Solvent consumption/kg of crude processed	130 dm^3
Annual production	36,000 kg

This is a relatively simple separation, so column loading is relatively high and solvent consumption per kilogram of crude material processed is low.

A sledgehammer to crack a nut!

The capital investment to establish a plant capable of handling a 1 m diameter column will be vast compared to the straightforward task of setting up a 15 cm diameter column in a laboratory.

It is important to consider all components of a purification when deciding which mode of chromatography to use for a particular separation. For example, when comparing normal phase with reversed phase chromatography the loading capacity of the latter is generally lower. However, the solvents used for reversed phase chromatography are predominantly aqueous and consequently are normally cheaper. The stationary phases used for reversed phase separations are, on average, twice the price of the normal phase equivalent but the resolving power is generally much greater. In some instances, as is the case with most biological molecules, recovery of the product from the aqueous eluent can be tedious. It can also be said that most biological molecules would not purify readily in purely organic solvents by normal phase chromatography.

5

Process development and optimization

Clearly it is not possible to describe every technique, or variation in a particular mode of chromatography, so this chapter will concentrate on the most dominant approaches to optimization of preparative HPLC. The objective here is to maximize the column load and to minimize mobile phase wastage allowing the purification of the largest amount of target in the minimum space.

Two case studies will be shown here to demonstrate the development of purification processes in both overload and resolution based separations[5.1]. The first example summarizes the purification of a synthetic peptide by overload chromatography, or more accurately described as sample self displacement chromatography. The techniques applied to this separation are applicable to any molecule and can be applied to all modes of chromatography, with the exception of size exclusion chromatography.

In the second example the separation of two optical isomers by batch chromatography using 'box car' injections demonstrates one approach to purification by chiral chromatography.

5.1 Sample Self Displacement for Purification of a Peptide

The purification of synthetic peptide objectives is open to many modes of separation due to the hydrophilic, hydrophobic and charged nature of amino acid side-chains. Employment of ion-exchange, hydrophilic interaction, normal phase and reversed phase modes of chromatography all seem perfectly viable. Though ion-exchange has its place in peptide purification, the complex and subtle nature of the impurities observed in peptides prepared by the solid phase approach dictates the use of high performance techniques offered by Normal and Reversed Phase modes of operation.

Synthetic peptides often contain impurities with no charge difference, with chemical modifications at side-chains, amino acid deletions and insertions, or epimeric impurities arising from racemization. This complexity has led to the development of many Reversed Phase media with carefully controlled particle size and pore size that result in high resolution under gradient elution. Many of these stationary phases are based on a silica matrix chemically bonded with alkyl chains such as C_8 or C_{18}. Though considerable effort has been made to mask the silica matrix, these supports suffer from an inherent base instability at levels above pH 9. In addition, in most cases the fragile nature of the silica matrix also reduces the lifetime of these media in high pressure systems. This is a particular problem for wide-pore, silica based stationary phases with a mean pore size of >200 Å and leads fracture of the media, generation of fines and eventual blockage of column frits. Conversely, for Reversed Phase chromatography, polymeric stationary phases offer both chemical and mechanical stability with none of the underlying traits of silica which is, by its very nature, a Normal Phase base matrix. Particle fragility and the secondary interactions caused by the cationic, hydrophilic nature of silica are not observed.

Peptides often have a low mobility in solution so a small particle size and uniform pore shape are essential for effective mass transfer. The tertiary structure of peptides differ dramatically from one molecule to the next so stationary phases that offer a range of pore sizes is prerequisite. Relatively small peptides containing 20–30 amino acids can often have a comparatively large radius of gyration in solution compared to some globular peptide structures containing significantly more amino acids. For instance, the radius of gyration of the Calcitonin peptides, which contain 32 amino acids, is ~20 Å whereas the diameter of Insulin, a 54 amino acid peptide, is ~15 Å. As a consequence, the Calcitonins show poor mass transfer in stationary phases with a mean pore diameter of 100 Å, whilst Insulin is readily, and usually, purified at high load on the same material. For an efficient and cost effective purification it is therefore essential to choose the appropriate surface characteristics, pore size and flow rate for each peptide.

For analytical separations it is important to run a full gradient based separation since this will recognize and allow subsequent identification of the majority of impurities present in the crude peptide. This impurity identification may eventually lead to improvements in the synthetic strategy by identifying specific shortfalls in the chemistry and ultimately to a higher yield coupled with lower manufacturing costs. Gradient elution is also particularly efficient when it comes to analysing low level impurities running close to the target peptide on the chromatogram that might otherwise be lost in the background noise of an isocratic separation.

In cases where mass transfer is rapid, as is the case with most small molecule separations, then isocratic elution can offer advantages such as automatic fraction reprocessing and solvent recycle. However, with larger synthetic objectives the rate of mass transfer is comparatively low so isocratic elution leads to band broadening and subsequently to recovery of the peptide at high dilution. Most preparative HPLC based peptide separations are carried out under gradient and overload conditions that allow for maximum throughput in terms of time and quantity.

Once the stationary phase and gradient conditions have been selected there are two options for practical applications. The first option is to maximize the column loading such that a reasonable level of resolution is maintained allowing peak skimming to achieve the required purity. The second option is to overload the column to such an extent that the main component displaces earlier running components of a mixture.

Whilst the former process will clearly yield material of the required purity it is not normally suitable for larger scale separations of synthetic peptides since column loadings will maximize at 1–2 mg/cm^3. In sample self-displacement chromatography the load is steadily increased until the main component displaces the less strongly bound impurity whilst minimizing the 'tag along' effect, of more strongly bound impurities, often observed with this mode of operation. In this approach the column loadings tend to maximize at 50–200 mg/cm^3, whilst the separation is effected in a similar time-scale. It is often beneficial to reverse the elution order of impurities, by a change of stationary phase, buffer, or pH, in order to allow self-displacement of these components of the mixture.

The self-displacement approach to purification of a synthetic peptide containing 25 amino acids is shown below[5.1]. The analytical HPLC trace (Figure 5.1) shows the impurity peak (A), which is displaced by the main component (B) in the demonstration.

The anticipated effect of column overload and eventual effect of displacement is shown diagrammatically in Figure 5.2.

The size of the 'blue' overlapping area is governed by the column efficiency. The more efficient the column, the smaller the overlap (Figure 5.3).

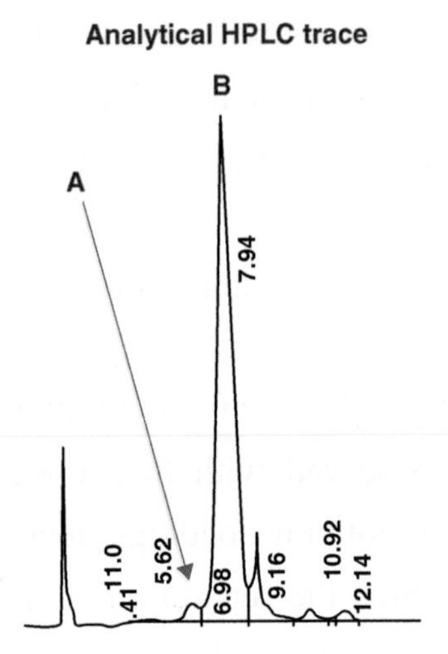

Figure 5.1 Analytical HPLC of synthetic peptide (25 amino acids)

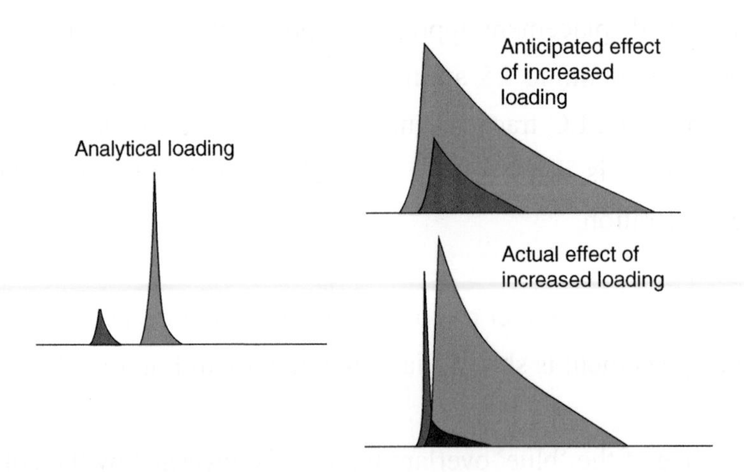

Figure 5.2 Observations during overload

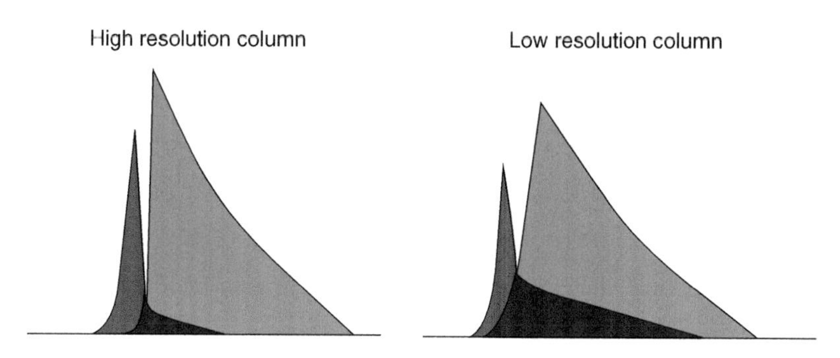

Figure 5.3

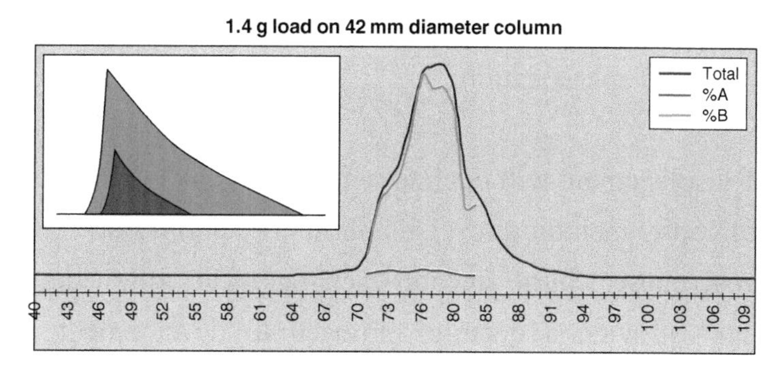

Figure 5.4

The effect of a comparatively low loading (1.4 g) on a semi-preparative column (42 m × 250 mm) packed with PLRP-S (8 μm, 300 Å pore) is shown in Figure 5.4. The blue line represents the actual UV trace obtained during the separation, the green line shows the relative purity of the main component (B) in relation to the impurity (A) and the red line represents the relative purity of the impurity (A) in relation to the main component (B).

It is clear that at this stage, the impurity is neither resolved or displaced. A gradual increase in loading eventually resulted in self-displacement and the optimum load for this particular molecule was ~6 g on the 42 mm diameter column (Figure 5.5). Again the blue line represents the UV trace obtained during the separation and the green and red lines represent the relative levels of main component and impurity, respectively. Self-displacement is characterized by the sharp increase in the level of impurity in the early running fractions whilst in this example the material collected between 89 and 130 minutes fell within the required specification.

Self-displacement will operate with almost any mode of chromatography including reversed phase as demonstrated above, normal phase, chiral and ion-exchange. The following procedures provide a generic approach to development of a

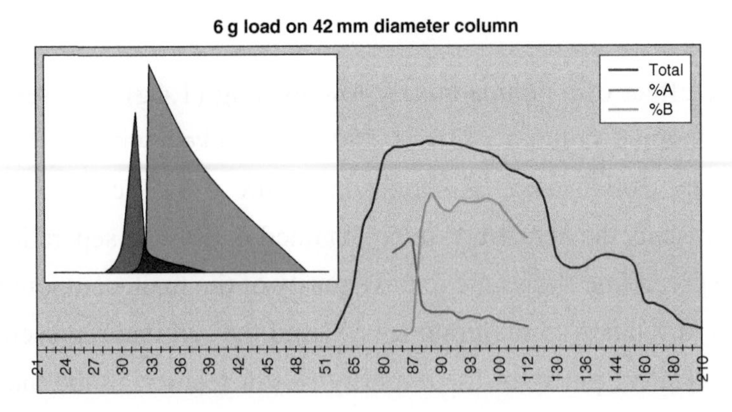

Figure 5.5

separation including choice of gradient, loading studies and flow rate optimization.

The first two stages can be performed on an analytical column. However, flow rate optimization is best performed on a column of at least 50 mm diameter in order to compensate for friction and column wall effects. The passage of solvent over particles in a packed HPLC column generates friction, which in turn produces heat resulting in a rise in temperature of the eluent. It is not uncommon with column lengths of 25 cm to see a temperature gradient of 2–3°C between inlet and outlet of the column. Though this does not sound phenomenal, when considered in conjunction with solvent cooling at the column walls, it can have a considerable effect on a chromatographic separation.

As the eluent passes through the column the heat generated reduces the viscosity of thc solvent. Cooling of the solvent at the column walls by conduction reduces viscosity in this vicinity resulting in a parabolic flow of eluent through the column, with solvent closer to the centre of the column moving faster than solvent at the periphery. There are several ways to reduce this effect, including heating the solvent in order to minimize the viscosity gradient and heating the column walls to reduce cooling by conduction.

The simplest and often the most cost effective way to combat friction is to reduce flow rate to a minimum. By no coincidence, this often leads to an increase in the efficiency of a separation since in many circumstances for preparative purifications, the less experienced have followed a linear scale-up from analytical column flow rates. In an ideal world each separation should, at some stage, involve a flow rate optimization. The fundamental principles behind this are discussed by JJ van Deemter[5.2] in what is probably the most cited paper in the history of chromatography. In summary, this suggests doing a graphical plot of separation efficiency versus flow rate and is particularly important for peptide purification where mass transport is comparatively slow. The van Deemter equation in simplified form can be represented as:

$$H = A + \frac{B}{\nu} + C\nu$$

where, H = height equivalent to a theoretical plate, and ν = flow rate.

Optimum flow rates are achieved at the lowest value for H. The terms A, B and C are constants relating to eddy diffusion, longitudinal diffusion and to mass transfer of the analyte.

The typical appearance of a van Deemter plot is shown in Figure 5.6. At low flow rates an axial longitudinal diffusion results in band broadening (the analyte can diffuse against the

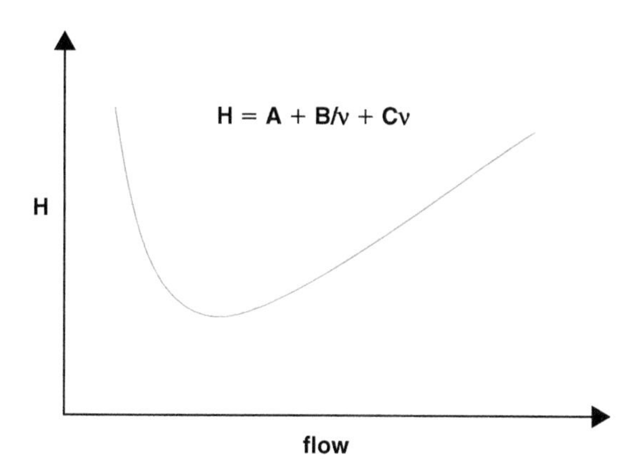

Figure 5.6

flow). Whilst at high flow rates the tortuous path followed by the eluent around the particles creates eddies, a resultant poor longitudinal diffusion and inefficient mass transfer lead to band broadening (in effect, the analyte cannot keep up with the solvent front).

The eluents most commonly used in peptide purification by reversed phase HPLC are water and acetonitrile. These are often buffered with trifluoroacetic acid (0.1% v/v, TFA), ammonium acetate (0.05–0.1 mol/dm^3 at pH 4–8) or phosphate (0.05–0.1 mol/dm^3 sodium or potassium salt at pH 2–8). In addition, polymeric reversed phase media also performs well at high pH and is often buffered with ammonium hydroxide or ammonium bicarbonate (0.05–0.1 mol/dm^3 at pH 8–9).

As stated above, it is useful to run an analytical gradient that recognizes all impurities in the crude peptide. In addition, this will identify the concentration of polar modifier at the point of desorption of the main component from the column. It is necessary to collect this data in order to use the generic approach to purification development discussed below. The analytical column used to select the preparative gradient elution conditions and in the loading study below should be packed with the stationary phase to be used for the preparative separation.

Analytical HPLC

Attach an analytical column (e.g. 4.6 mm × 250 mm) to the HPLC system and program the equipment to run the gradient summarized in Table 5.1.

Table 5.1

Time (minutes)	Concentration of buffer B (% v/v)	Flow rate (cm^3/min)
0	1	1.00
30	100	1.00
33	100	1.00
36	1	1.00
40	1	1.00

Dissolve the crude peptide (1 mg) in the starting buffer ($1\ cm^3$), inject the solution ($20\ mm^3$) and elute using the gradient shown in Table 5.1 analysing at a wavelength of 220 nm for peptides.

Compensating for the column void volume, calculate the concentration (of buffer B) at which the main component is desorbed.

Preparative HPLC can be carried out in pre-packed HPLC columns or self-pack systems (e.g. DAC columns). The conditions used to slurry pack various self-pack hardware is discussed in Chapter 4.

Gradient selection

In order to select the elution conditions for a preparative separation, the concentration of polar modifier at the point of desorption is required (see above). As a general rule of thumb, under overload conditions the main component will elute at approximately two-thirds of the concentration of polar modifier observed in the analytical separation. For example, if a main component desorbs from the analytical separation at a concentration of 30% B under overload conditions it will elute at approximately 20% B.

Table 5.2

Time (minutes)	Concentration of buffer B (% v/v)	Flow rate (cm^3/min)
0	10	0.50
60	30	0.50
70	100	0.50
80	10	0.50

A generic gradient recommended for preparative separations is to start the elution at a concentration of B that corresponds to a value 10% earlier that the anticipated desorption then apply a change in concentration of B corresponding to 5% over 15 minutes. To exemplify this approach Table 5.2 shows a gradient, applied to an analytical column, used to purify a peptide that is anticipated to desorb at 20% B under overload conditions.

It is important to note that the preparative separation is carried out at a low flow rate to allow for mass transfer, diffusion and column wall effects discussed previously.

Loading study

Attach an analytical column (e.g. 4.6 mm × 250 mm) to the HPLC system and program the equipment to run the chosen preparative gradient. Run a series of separations at column loadings of 40, 60, 80, 100, 120, 140 and 160 mg collecting 0.5 minute fractions during each elution. Analyse these fractions to determine the optimum load for self-displacement of early running impurities.

Flow rate optimization

For 'one off' separations flow rate optimization is not strictly necessary. However, it does offer an additional level of optimization for long-term projects and ongoing manufacture. The ideal flow rate often determined for small molecules on stationary phases with 10 μm particle size and 60–300 Å pores is approximately 120 cm/h, where for the majority of peptides and proteins the figure is typically in the region of 100 cm/h.

In order to accumulate the data for a van Deemter plot (described above) it is necessary to carry out a series of separations on a preparative column of at least 50 mm diameter. This can be done isocratically or using the gradient and optimum load conditions selected above. However, if the study is done using gradient elution the gradient length should be reduced in proportion to increased flow rate. Suggested linear flow rates for this study are 60, 75, 90, 105, 120 and 135 cm/h. These flow rates correspond to approximately 20, 25, 30, 35, 40 and 45 cm^3/min for a 50 mm diameter column.

Although it is not strictly valid to determine column efficiency using plate count theory under gradient elution the actual benefit in measuring the 'effective plate height', H in this exercise will be validated by an improvement in the separation efficiency. This can be explained by imagining a finite point in the gradient. At this finite point the mechanism of desorption is independent of flow rate and is completely controlled by the eluent composition, which is always constant at that point in time. However, the mechanics of diffusion and mass transfer are still dependent on the flow rate.

Calculate the effective number of theoretical plates, N, and subsequently the value of H for each separation using the following formulae:

$$N = 5.54\left(\frac{t}{w_{0.5}}\right)^2$$

where t = time of elution, and $w_{0.5}$ = peak width at half height

$$H = \frac{L}{N}$$

where L = column length.

Prepare a plot of H versus flow rate. The graph should resemble that shown in Figure 5.6 and the optimum flow rate is determined as the lowest point in the valley. If the main component has 'saturated' the detector, a close running peak can be used for the calculations.

Table 5.3

Time (minutes)	Concentration of buffer B (% v/v)	Flow rate (cm^3/min)
0	10	33
60	30	33
70	100	33
80	10	33

Attach the preparative column (e.g. 50 mm × 250 mm) to the HPLC system and program the equipment to run the gradient conditions determined above. A typical gradient, where the main component desorbs at 30% B on the analytical separation and the optimum flow is determined to be 100 cm/h, is exemplified in Table 5.3 for purification on a 50 mm diameter column.

5.2 Boxcar Injections for Chiral Separations

As described in Chapter 1, the optimum process for separation of a binary mixture is more than often SMB chromatography due to the continuous nature and the efficient use of both mobile phase and stationary phase. However, the effort required to both optimize and stabilize a SMB based separation is complex and time consuming. Consequently, for 'one off' separations it may not be prudent to use a technique that is better suited to established industrial processes. In addition, in most circumstances the expensive equipment required for SMB chromatography may not be readily available.

The technique of 'boxcar injections' (not to be confused with boxcar chromatography) can be extremely productive for isocratic elution in any mode of chromatography and should always be considered when scaling up a separation. The preparative HPLC of an enantiomeric mixture utilising a chiral stationary phase is described here to demonstrate the approach for separation of a binary mixture.

For this example a loading study, similar to that described above, performed on the crude mixture suggested that a good separation would be achieved at loadings equivalent to 20 mg on an analytical column (4.6 mm × 250 mm). The required component of the mixture, shown on the analytical HPLC trace in Figure 5.7, was the more retained peak.

The equivalent load of 8 g on a 10 cm diameter column was actually shown to give even better resolution (Figure 5.8). However, for this particular project this loading was still too low to give the required productivity so the effect of increased loading was investigated further.

At a loading of 20 g on the 10 cm diameter column a third peak was observed which, upon further investigation, proved to be

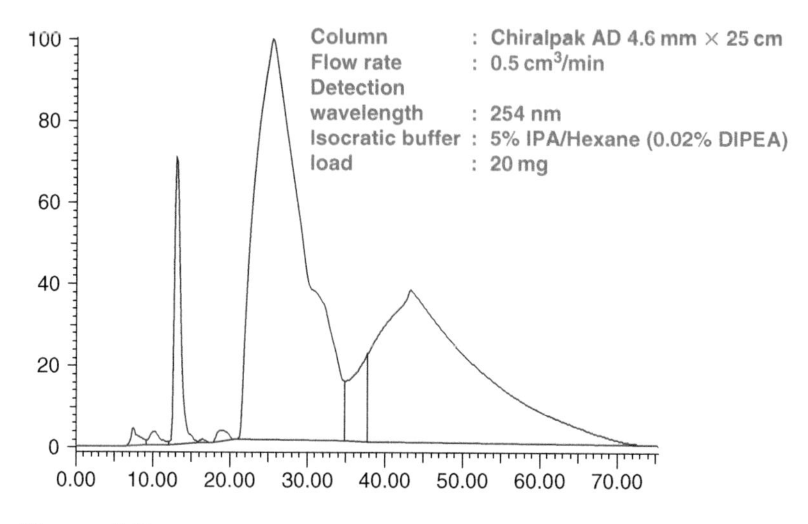

Figure 5.7

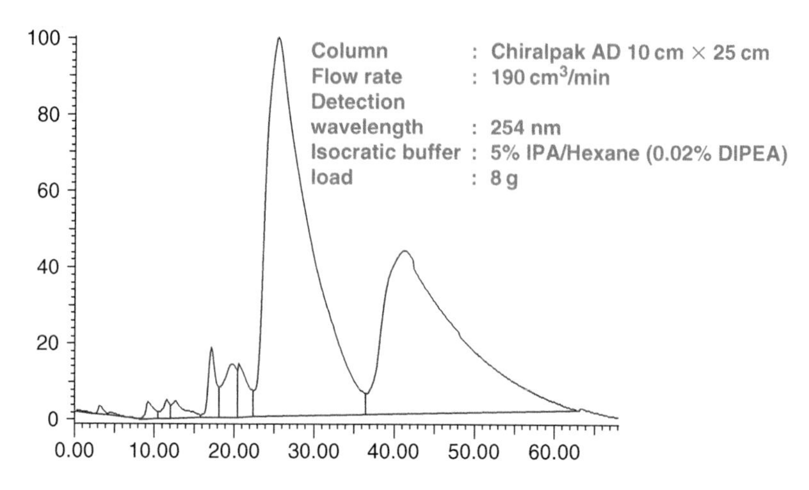

Figure 5.8

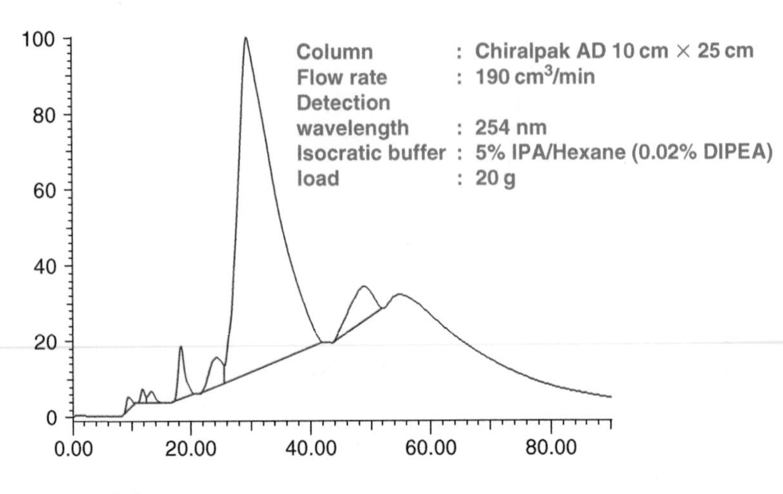

Figure 5.9

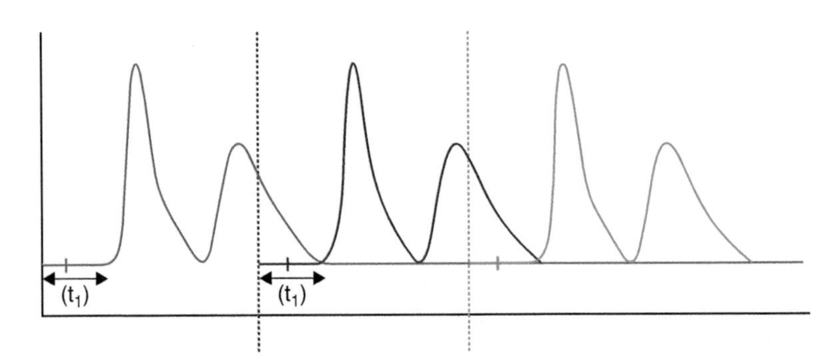

Figure 5.10

due to sample self displacement of the tale of the early eluting component by the more strongly retained material (Figure 5.9).

In order to increase productivity it was decided to use 'boxcar injections' to isolate the required component. This process, described diagrammatically in Figure 5.10, utilizes repeat

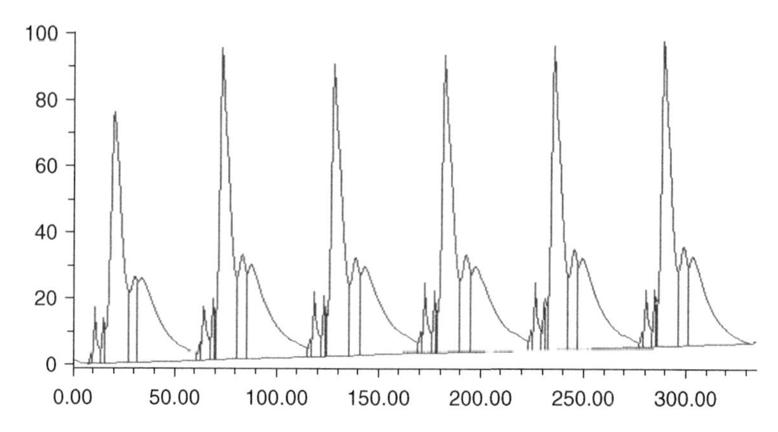

Figure 5.11

injections in rapid succession to maximize throughput. The time at which the first peak is eluted (t_1) is subtracted from the time that elution of the last peak is complete. The second injection is made at this time point and the process repeated as many times as necessary.

In this particular example six simultaneous injections at 20 g load were performed in less than 6 hours (Figure 5.11).

It is important to note that the technique of boxcar injections is by no means restricted to chiral separations and should always be considered for any separation performed under isocratic elution.

6

Documentation and record keeping

Chapter Six

A large number of the companies involved in preparative HPLC will be using the technique for purification of drugs to cGMP. This chapter will summarize some of the documentation required including equipment qualifications, standard operating procedures, process instructions and process records. It will give some suggested layouts for process documentation with real examples for guidance.

Regulatory guidelines and nomenclature will differ from one country to the next but the general principle remains the same. The Federal Drug Agency (FDA), which is undoubtedly the most discerning authority, presides in the United States. The *Orange Guide* published by the MHRA[6.1] is useful text and to some extent shows a level of harmonization of the guidelines with the FDA.

6.1 Equipment Qualification

The principle for the documentation trail and qualification process surrounding the purchase installation of equipment is a matter of common sense. The following terms are by no means universal but the principle and elements of these are essential to meeting cGMP.

User Requirement Specification (URS) is a basic description of the equipment covering the purpose for which it is required.

This is sent to equipment suppliers during the stage of assessing various manufacturers and obtaining cost estimates.

Design Qualification (DQ) is a much more detailed document describing the requirements in full. The DQ may contain engineering drawings and performance specifications such as permitted variance on gradient or flow rate. This document will specify precise requirements such as the grade of stainless steel for construction and the level of polish on the interior walls of a column.

Factory Acceptance Tests (FATs) are performed at the site of manufacture of a piece of equipment before it is shipped to the customer. These ensure that the equipment meets the DQ precisely.

Site Acceptance Tests (SATs) generally carried out by the site installation engineers in conjunction with the manufacturer, ensures that the equipment is installed appropriately and still meets the requirements of the FATs. Any damage occurring before or during shipping will be spotted at this point.

Installation Qualification (IQ) is a formal document confirming that the equipment meets the URS and SATs. Though this may contain elements of the SATs this document will now contain inspections and approvals from the Quality Assurance (QA) department and the Technical Management for the facility.

Operational Qualification (OQ) is a QA approved document designed to check that the individual components operate according to specification. Pumps will be tested with calibrated flow meters to confirm that they perform to the set specification. Calibration standards will be used to assess other instrumentation such as Refractive Index (RI) or Ultra Violet (UV) detectors. Gradient performance will be checked to confirm that the solvent delivery system and mixer is working to specification. Once complete and approved by QA and Technical Management, the equipment will now be made available for use.

Performance Qualification (PQ) is an ongoing process where the equipment is routinely tested to ensure that it is still performing to the standards met in the OQ. Guidelines differ depending upon the governing authority but bi-annual or annual checks are commonplace.

6.2 Process Documentation

Standard Operating Procedures (SOPs) are prerequisite for any process involving the preparation of an Active Pharmaceutical Ingredient (API) and effectively provide an instruction of what to do. With reference to preparative HPLC there will be an SOP for each step and this may stretch back as far as an SOP to describe the format and units used in writing an SOP.

In more practical terms there will be SOPs to describe sample preparation, equipment set-up including programming, and an SOP for operation of the equipment. The documentation does not stop here since there will be defined instructions for fraction analysis, fraction reconciliation and recovery of product from the fractions.

Process Instructions (PIs) and Process Records (PRs) are batch specific documents used to follow the progress of each individual stage. As mentioned previously these terms are not universal and these documents are often referred to as Batch Manufacturing Instructions (BMIs) and Batch Manufacturing Records (BMRs). Typical layouts and examples of PIs and PRs are given in the appendices – it is clear from the appendices that there is a major emphasis on approval and witnessing.

The first page of the PI is approved by a hierarchy of personnel starting with the QA department which is responsible for

Does the development have to be cGMP compliant?

Some authorities will insist on inspecting the documentation trail for process development, including laboratory notebooks.

ensuring that the content and format of the document mccts the required quality standards for the appropriate authorities. The Development Manager is responsible for ensuring that the documents will allow the process to be carried out as it was originally developed.

The Senior Chemist in charge of the process is signing to say that he or she understands the process and will ensure that the documentation is completed. The Technical Manager, shown in the examples as the GMP Facilities Manager, will take overall responsibility for ensuring that the documentation is suitable for purpose and ultimately, that it is completed correctly. Although the manager and operators are responsible for carrying out the process safely the documentation is often additionally approved by a safety representative.

In the example given, the second page lists the various stages of a process starting with the preparation stage. It is essential that the abbreviations used throughout are consistent so guidelines will be described or referred to here. It is common practice to use the conventions agreed by the International Union for Pure and Applied Chemists (IUPAC) and similarly to employ SI units for weights and measures.

In the case of purification there will be target specifications for the final product and these must be clearly defined at the start of the process. The raw material specifications are equally important since it will be impossible to reproduce a separation from batch to batch if the feedstock and reagents are not consistent. The equipment used and origin of consumables will also be documented.

Quality checks on consumables!

If these have contact with the product it is necessary to determine that they are fit for purpose. Plastic pipettes, for instance contain leachables that could contaminate a product.

In-process tests will appear at critical stages in a process to ensure all is going to plan and that any deviations from the set process can be explained. The SOPs required to support the process will be listed so that they can be made available at the time of use. Since a batch of crude product will be purified in several portions there will be some form of cleaning between operations. This may include anything from washing the solvent lines to a fraction collector through to regeneration of the stationary phase.

Before the operation can proceed, the personnel involved will be given a list of responsibilities and in most countries the material safety data sheets (MSDSs) will be made available.

The process now begins and the PRs appropriate for each stage must be completed at the time of each stage of the operation. At certain critical stages, such as fraction pooling, a witness or managerial approval may be required. Deviations must be recorded on the appropriate Deviation Record.

Don't lose a sheet!

Number each sheet of each document 'Page ..... of ....'

Appendices

PROCESS INSTRUCTIONS

STAGE 3 – PREPARATIVE HPLC

PI Number	Modification Number	Job Number	Product Code Number
PI 0048	1	14449	DW-12-14449

APPROVAL

	Signature	Date
Quality Assurance	______________	______________
Development Manager	______________	______________
Senior Chemist	______________	______________
GMP Facilities Manager	______________	______________
Safety	______________	______________

CONTENTS

DEVIATION SHEET

This sheet shall be used to summarize any deviations that took place during the process. The deviation shall include reference to the specific step/steps and the corresponding Deviation Reference Number.

NOTE: All deviations must be agreed in advance by the GMP Facilities Manager before proceeding.

OPERATION 1 PREPARATION SECTION

1.01 Abbreviations

Abbreviations used in this document are consistent with those recommended by the International Union of Pure and Applied Chemists (IUPAC) unless otherwise stated.

Abbreviation	**IUPAC nomenclature**
TFA	Trifluoroacctic acid

1.02 Target Specifications

To achieve ⩾97% purity, with no single impurity >1.5%, except up to 10% dimer as determined by the reversed phase HPLC conditions set out in **1.06**. The purification is to be achieved by Reversed Phase Preparative HPLC using a 110 mm DAC column packed with 10 μm C8 100 Å to a length of approximately 25 cm. The product will be purified by gradient elution using the 'Guideline Purification Strategy' attached.

1.03 Raw Materials

Material	**RM SOP Number**	**Approved for Use (QC cleared, within expiry date)**
TFA	RM 1041	
Acetone	RM 1004	
MeCN	RM 1006	
C8 10 μm 100 Å	RM 1056	
RO water	N/A	

1.04 Equipment

Item	ES SOP Number	Initials and Date	Serial Number
2 × Rainin SD-1 Pumps, fitted with 800 cm^3 pump heads**	ES0516		
Rainin UV-1 UV/Vis Detector**	ES0518		
Dynamax PC Chromatography Data System	ES0429		
Prochrom 110 mm column**	ES0385		
Duran bottles 20 dm^3			
Waters 600 Pump	ES0342		
Waters 481 Detector	ES0343		
Waters 740 Integrator	ES0344		
Waters 712 Wisp	ES0345		
Millipore glass filter holder (1 dm^3)	ES0425		
Glass microfibre filter unit*	N/A		
Top Pan Balance	ES0300		To be recorded when used
Conical Flask*			
Measuring Cylinders**			
Gilson Precision Pipette, 50–200 mm^3	ES0030		

The equipment SOPs should be read prior to starting this stage of the process.

*Note these items shall be dedicated to this product.

**Note these items shall be dedicated to purification, labelled as such and stored when not in use. Cleaning Verification shall be performed between projects for all product contact surfaces as appropriate.

Serial numbers shall be added to the above table when the item is used. For critical items the serial numbers shall be recorded on the process record where indicated at the point of use.

1.05 Consumables

Item	RM SOP Number
HPLC vials	
0.5 dm^3 Duran bottles	
Pasteur pipettes	
Whatman glass microfibre filters	
Analytical column	

1.06 In-Process Tests

The progress of the purification will be monitored by analytical HPLC. See the analytical HPLC gradient for details. This method will have been developed in the Development Group and refined in the QC department

Analytical HPLC Conditions

Column: C_{18} 250 × 4.6 mm

Sample: Fractions

Buffer A: 0.1% TFA in water

Buffer B: 60% MeCN in Buffer A

Flow rate: 1.0 $cm^3 \cdot min^{-1}$ $\lambda = 210$ nm

Temperature Ambient

Gradient: 0–100% B over 40 minutes

Analytical HPLC Strength Method

Various purification fractions or fraction pools may be analysed for strength by peak area on HPLC. The analytical HPLC conditions shown above will be used. A standard sample with a known product content will be supplied by the Development Group. The peak area on HPLC of a solution of the standard (1.0 mg/cm^3, net) will be compared with the percentage area of the sample under test to estimate the quantity of the net product present.

1.07 Ancillary SOPs

SOP Title	SOP Number	Initials and Date
Preparation and use of the HPLC Column Standard	QC0217	
Preparation of Buffers containing TFA (0.1% v/v) for Analytical HPLC	SP0511	
Preparation of MeCN in Water (1:1 v/v)	SP0742	
Column cleaning gradients	SP0650	
Assembly and packing of the Prochrom column	SP0514	
Recovery of Product from Large Scale Purification	SP0506	

1.08 Cleaning

All glassware used in the purification shall be project dedicated and appropriately labelled for the project. New glassware will be rinsed with MilliQ H_2O prior to a final wash with acetone before use. All glassware that is repeatedly used will be cleaned using the guidelines set out in the Cleaning of Laboratory Glassware SOP (SPO523) and this shall be recorded prior to the commencement of each operation. Where applicable Cleaning Verification Reports must be cross referenced.

1.09 Personnel and Responsibilities

Name	Responsibility	Signature	Initials and Date
Joe Bloggs	Senior Scientist with overall responsibility for the purification process		
Fred Soap	Production Chemist assisting with the purification process and checking steps as required		
Don Wellings	GMP Facilities Manager responsible for checking that the purification process has been carried out correctly and agreeing any deviations from the process		

1.10 Safety

The following COSHH, Hazardous Experimental Assessments and Material Safety Data Sheets (MSDS) should be read before starting this process. Sign and date the table below after reading the assessments and MSDSs.

HAZARDOUS ASSESSMENT	REFERENCE CODE	Initials and Date
Solvent Preparation for Preparative Chromatography	COSHHA008	
Sample Recovery from Preparative Chromatography	COSHHA009	
Lyophilization of Peptides	COSHHA010	
Packing a Preparative Chromatography Column	COSHHA011	
Sample Preparation for Preparative Chromatography	COSHHA016	
Preparative HPLC	COSHHA01	

COSHH ASSESSMENT	REFERENCE CODE	Initials and Date
Preparation of Samples for Purifications	COSHH024	
Loading of Samples for Purification	COSHHA026	
Fraction Collection	COSHHA027	
Preparation of Samples for HPLC Analysis	COSHHA029	
Analytical HPLC	COSHHA032	
Pooling of HPLC Fractions	COSHHA033	
Washing Glassware using Acetone	COSHHA037	
Use of 70% propan-2-ol for Cleaning Laboratories	COSHHA074	
Disposal of Chemicals	COSHHA076	
Transport of Chemicals	COSHHA077	

Safety Precautions

Safety glasses and coverall suits/lab coats must be worn at all times by personnel in the laboratory.

Whilst handling all chemicals, gloves must be worn.

When TFA is handled gauntlets or arm/sleeve protectors will be used.

All procedures should be carried out within a fume cupboard, apart from weighing which can be undertaken in a ventilated cabinet. Prior to using the ventilated cabinet the operator should check that the LEV is working.

Ensure that the fume cupboard is as tidy as possible, that the extraction is working efficiently and that only one 'port' of the fume cupboard is open at any one time (an alarm will go off if the face velocity drops below the minimum safe operating level) or the fume cupboard sash is at its lowest practical level when using the older fume cupboards. When the operator is absent all the 'ports'/the sash should be closed.

Check that the fume cupboards are adequately bonded.

Check that any electrical appliance to be used has been passed by an electrician. If it has been passed by the electrician and appears to be faulty get it rechecked before use.

Any spillage should be dealt with quickly and with reference to the substances COSHH assessment or MSDS. BA equipment may be required. Other working areas may need informing as the laboratory is at positive pressure to all surrounding areas.

If whilst handling the chemicals involved in these processes, contact is made with your skin, eyes, by inhalation or by ingestion, seek medical help initially from a First Aider. Refer to the MSDS

All glassware should be checked for cracks and chips prior to use.

All accidents and spillages should be reported to your supervisor and safety officer.

The Project 'HazOp' Study should be referred to for assessment of the risks associated with this project.

OPERATION 2 PRE-PURIFICATION COLUMN PREPARATION

For all steps record any observations and minor deviations on the blank section alongside the process prompts. This applies for all the subsequent Operations.

Manufacturing Facility ________ **Date** ________

2.01 Confirm that the DAC Column is packed with C_8 (2.0 kg of 10 μm, 100 Å pore size) to achieve a minimum 25 cm column length. Record the following:

GI number of C_8 ________
Packing operation number ________
Operator ________
Date ________

2.02 Prepare a solution of Buffer A (0.1% v/vTFA in H_2O) and Buffer B (0.1% v/vTFA in MeCN) for use in purification. Record the following:

Batch number of Buffer A ________
Batch number of Buffer B ________
Operator ________
Date ________

OPERATION 3 PURIFICATION P____

Manufacturing Facility __________ Date __________

For all steps record any observations and minor deviations on the blank section alongside the process prompts. This applies for all the subsequent Operations.

3.01 Prior to this operation clean glassware to be cleaned as indicated in SOP SPO523. Indicate that cleaning is completed as appropriate:

4 dm^3 conical flask __________
250 cm^3 measuring cylinder __________
1 dm^3 measuring cylinder __________
Magnetic follower __________
Glass microfibre filter unit __________
5 dm^3 filter flask __________
Membrane filter holder __________
Operator __________
Date __________

3.02 Power up the preparative HPLC system. Set the detector wavelength to 230 nm. Equilibrate the column in 8% Buffer B at $190\,cm^3 \cdot min^{-1}$ for 22 min. This time is approximately twice the column volume (column volume is ~$2000\,cm^3$). Record the following:

Time equilibration started ________

Time equilibration completed ________

Equilibration time ________

Column compression pressure ________

Operator ________

Date ________

Checked by ________

Date ________

3.03 Weigh 50 g gross weight of crude product (record the calculation below) into a $4\,dm^3$ conical flask. Add a magnetic stirrer follower and place on a magnetic stirrer. Record the following:

Weight of flask ________

Weight of flask + product ________

Weight of product ________

Sample number of product ________

Serial number of balance ________

Operator ________

Date ________

Checked by ________

Date ________

3.04 Add 250 cm^3 of Buffer B to the crude product and start stirring. Record the following:

Batch number of Buffer B __________
Volume of Buffer B __________
Operator __________
Date __________

3.05 Add 2.25 dm^3 of Buffer A slowly over 10 minutes and stir until all the solid has dissolved, approximately 30 minutes. Record the following:

Time stirring started __________
Time stirring complete __________
Elapsed time for stirring __________
Batch number of Buffer A __________
Volume of Buffer A __________
Operator __________
Date __________

3.06 Filter the crude product solution through a GF/D glass microfibre filter into a 4 dm^3 filter flask. Rinse the 4 dm^3 conical flask with $3 \times 50\ cm^3$ of Buffer B in Buffer A (8% v/v) and transfer to the filter. Record the following:

GI number of filter membrane ________
Volume of Buffer B in Buffer A (8% v/v) ________
Batch number of Buffer B in Buffer A (8% v/v) ________
Number of rinses ________
Volume of rinses ________
Operator ________
Date ________

3.07 Filter the crude product solution through a GF/B filter into a 4 dm^3 filter flask. Rinse the 4 dm^3 conical flask from step **3.06** with $3 \times 50\ cm^3$ of Buffer B in Buffer A (8% v/v) and transfer to the filter. Remove a sample for analytical HPLC refer to 1.06, label the sample 14449/0048/3.07(s), attach the trace to this process record. Record the following:

GI number of filter membrane ________
Volume of Buffer B in Buffer A (8% v/v) ________
Batch number of Buffer B in Buffer A (8% v/v) ________
Number of rinses ________
Volume of rinses ________
Operator ________
Date ________

3.08 Restart the flow on the HPLC in the equilibration conditions. Zero the UV/Vis detector. Load the crude product solution onto the column at $190\,cm^3 \cdot min^{-1}$ through Pump A. All eluent is to be collected. Record the following:

Time loading started _______
Time loading finished _______
Duration of loading _______
Detector zeroed (yes/no) _______
Operator _______
Date _______

3.09 After loading the crude product rinse out the flask from step **3.07** and injection line with $100\,cm^3$ of Buffer B in Buffer A (8% v/v). Record the following:

Volume of Buffer B in Buffer A (8% v/v) _______
Batch number of Buffer B in Buffer A (8% v/v) _______
Operator _______
Date _______

3.10 Run the column with 8% B at $190\,cm^3 \cdot min^{-1}$ for 5 minutes. Record the following:

Time solvent flow restarted _______
Time re-equilibration completed _______
Duration of re-equilibration _______
Operator _______
Date _______

3.11 Start the purification gradient. Start collecting 1 minute fractions manually when a significant UV response occurs. Continue collecting until the product has eluted. Record the time the gradient and fraction collection started. Label the fractions PX/FY-Z where Y and Z are sequential fraction numbers, X is the sequential purification number.

Time gradient started __________
Time fraction collection started __________
Duration of fraction collection __________
Number of fractions collected __________
Operator __________
Date __________

3.12 After the gradient has finished, remove samples for HPLC analysis (see 3.13). Cap the bottles and store fractions at 2–8°C (in laboratory fridge).

Time gradient finished __________
Duration of gradient __________
Storage location __________
Operator __________
Date __________

3.13 Record which fractions are to be analysed. Analyse these fractions by HPLC using the analytical HPLC conditions in Step 1.06. Attach all HPLC traces to this record.

List fractions to be analysed below.

Note: Dilute fractions X10 before analysis

Operator _________

Date _________

3.14 Prepare a trial pool in a snap cap vial consisting of 50 mm^3 of each fraction that meets the required specification by HPLC. Label the pool PX/FY-Z, where X is the purification number, Y and Z covers the fraction range in the pool. Dilute pool 10 times and carry out a HPLC of this pool to confirm that the pool meets the required specification. Record the following:

Sample number of pool _________

Does the pool meet specification (yes/no) _________

Operator _________

Date _________

3.15 If the trial pool does not meet the specification the Senior Chemist will use his/her judgement to assess a second trial pool. Record the details of further pools on a purification deviation sheet and attach to this process instruction.

Additional pools required (yes/no) __________

Operator __________

Date __________

PURIFICATION DEVIATION SHEET

This sheet shall be used to summarize additional trial pools of purification fractions for Step **3.15**.

Prepare a trial pool in a snap cap vial consisting of 50 mm^3 of each fraction that meets the required specification by HPLC. Label the pool PX/FY-Z, where X is the purification number and Y to Z covers the fraction range in the pool. Dilute the pool 10 times and carry out a HPLC of this pool to confirm that the pool meets the required specification. Attach the HPLC traces to this process instruction. Record the following:

Sample number of pool __________
Pool meets specification (yes/no) __________
HPLC of pool attached (yes/no) __________
Operator __________
Date __________

Prepare a trial pool in a snap cap vial consisting of 50 mm^3 of each fraction that meets the required specification by HPLC. Label the pool PX/FY-Z. Dilute pool 10 times and carry out a HPLC of this pool to confirm that the pool meets the required specification. Attach the HPLC traces to this process instruction. Record the following:

Sample number of pool __________
Pool meets specification (yes/no) __________
HPLC of pool attached (yes/no) __________
Operator __________
Date __________

Prepare a trial pool in a snap cap vial consisting of 50 mm^3 of each fraction that meets the required specification by HPLC. Label the pool PX/FY-Z. Dilute pool 10 times and carry out a HPLC of this pool to confirm that the pool meets the required specification. Attach the HPLC traces to this process instruction. Record the following:

Sample number of pool __________
Pool meets specification (yes/no) __________
HPLC of pool attached (yes/no) __________
Operator __________
Date __________

3.16 Determine the quantity of product in the trial pool using the following equation.

$$\frac{10 \times \text{Area of pooled fraction}}{\text{Area of strength standard}} \times 1.0\,\text{mg} \cdot \text{cm}^{-3}$$

$$= \text{concentration of product in pool/mg} \cdot \text{cm}^{-3}$$

$$\begin{bmatrix}\text{Concentration}\\ \text{of product}\\ (\text{mg} \cdot \text{cm}^{-3})\end{bmatrix} \times \begin{bmatrix}\text{volume}\\ \text{of pool}\\ (\text{cm}^{-3})\end{bmatrix} = \begin{bmatrix}\text{quantity of}\\ \text{product/mg}\end{bmatrix}$$

Reference Number of $1.0\,\text{mg} \cdot \text{cm}^{-3}$ standard __________

Area of pool __________

Area of standard __________

Concentration of product in pool ($\text{mg} \cdot \text{cm}^{-3}$) __________

Volume of pool __________

Quantity of product (mg) __________

Operator __________

Date __________

Checked by __________

Date __________

3.17 Record the fate of all fractions on the process record for Reconciliation of Fractions.

3.18 Store all fractions in the fridge until required for final freeze drying. Record the following:

Identity of fridge __________
Operator __________
Date __________

3.19 At the end of each day wash the purification system including the solvent lines, injection line, pumps, flow cell and column with MeCN/H_2O (5 dm^3,1:1 v/v) at 200 cm^3/min.

Volume of MeCN/H_2O __________
Batch number of MeCN/H_2O __________
Operator __________
Date __________

APPROVAL OF STAGE

GMP Facilities Manager		**Date:**
Quality Assurance		**Date:**

EXAMPLE OF FRACTION RECONCILIATION DOCUMENT

RECONCILIATION OF FRACTIONS (for Stage 5)

PI Number	Modification Number	Job Number	Product Code Number
PI0050	0	14506	BP-12-14231

APPROVALS	Signature	Date
Quality Assurance	____________	________
Senior Chemist	____________	________
Peptide Technology and GMP Facilities Manager	____________	________

OPERATION 1 SUMMARY

1.01 This process instruction is used to summarize the fate of fractions from up to four purifications carried out according to PI0043 (prior to the desalt process).

The fate of fractions that are not combined for desalt will also be retrospectively recorded on this process instruction.

Record the fate of all fractions on the reconciliation tables (Tables A1 to A3).

Date when all fraction reconciliation is complete ______

Operator ______

Senior Chemist ______

1.02 Record any additional notes on fraction reconciliation below.

Operator ________

Date ________

1.03 PURIFICATION NUMBER ____

Record any additional notes on table (e.g. net product from trial pool or actual pool)

Table A1

Starting Material	Fraction or fractions pooled	Total volume/dm^3	Net product content/g	Fate of pool or fractions	Initials and date

1.04 PURIFICATION NUMBER ____

Record any additional notes on table (e.g. net product from trial pool or actual pool)

Table A2

Starting Material	Fraction or fractions pooled	Total volume/dm^3	Net product content/g	Fate of pool or fractions	Initials and date

1.05 PURIFICATION NUMBER ____

Record any additional notes on table (e.g. net product from trial pool or actual pool)

Table A3

Starting Material	Fraction or fractions pooled	Total volume/dm^3	Net product content/g	Fate of pool or fractions	Initials and date

References

References

CHAPTER 1

[1.1] Tswett, M.S. *Botanisches Centralblatt.* **89**, 120–123 (1902).

[1.2] Couillard, F. *Chromatography Apparatus*, US Patent 4,597,866, 1986-07-01.

[1.3] Petro, M. and Nguyen, S.H. *Liquid Chromatography Column Distributor*. US Patent Application US20020099452, 3003-09-18.

[1.4] Colvin jr, A.E. and Hanley, M.W. *Fluid Control Device*. US Patent 4,894,152, 1990-01-16.

[1.5] McNeil, R. *Fluid Flow Control Device*. US Patent 4,354,932, 1982-10-19.

[1.6] Bayer Technologies.

[1.7] LePlang, M. and Chabrol, D. *Fluid Distributor and Device for Treating a Fluid such as a Chromatograph Equipped with said Distributor*. US Patent 5,141,635, 1992-08-05.

[1.8] Heuer, C., Hugo, P., Mann, G., and Seidel-Morgenstern, A. *Journal of Chromatography. A*, (1996) 19–29.

[1.9] Dapremont, O., Cox, G.B., Martin, M., Hilaireau, P. and Colin, H. *Journal of Chromatography. A*, (1998) 81–99.

[1.10] Cox, G.B. *Journal of Chromatography. A*, 599 (1992) 195–203.

[1.11] Nicoud, R-M. and Charton, F. *Journal of Chromatography. A*, 702 (1995) 97–112.

[1.12] Broughton, D.B. and Gerholg, C.G. US Patent (1961) 2,985-589.

[1.13] Barker, P.E. *Chem. Eng. Sci*. (1960) **13**, 82.

[1.14] Guest, D.W. *Journal of Chromatography. A*, 760 (1997) 159–163.

[1.15] Nicoud, R-M. *Separation Science and Technology*, (2000) **2**, 476–509.

[1.16] Gottsclich, N. and Kasche, V. *Journal of Chromatography. A*, (1997) 201–206.

[1.17] Abel, S., Bäbler, M., Arpagaus, C., Mazzotti, M. and Stadler, J. *Journal of Chromatography. A*, (2004) 201–210.

CHAPTER 2

[2.1] Snyder, R.L., Glajch, J.L. and Kirkland, J.J. *Practical HPLC Method Development.* John Wiley & Sons Inc. 1988. ISBN 0-471-62782-8.

[2.2] Brown, P.R. and Hartwick, R.A. *High Performance Liquid Chromatography.* John Wiley & Sons Inc. 1989. ISBN 0-471-84506-X.

[2.3] Martin, A.J.P. and Synge, R.L.M. *Biochem. J.* (1941) **35**, 91.

[2.4] van Deemter, J.J., Zuiderweg, F.J. and Klinkenberg, A. *Chem. Eng. Sci.*, (1956) **5**, 271–289.

CHAPTER 3

[3.1] Tswett, M.S. *Botanisches Centralblatt* (1902) **89**, 120–123.

[3.2] Wellings, D.A. *Stepwise Elaboration of Quasi-Homogeneous Peptide-Gel Networks*. PhD thesis submitted to the Council for National Academic Awards (1986).

[3.3] Pirkle, W.H. and Anderson, R.W. *J. Org. Chem.* (1974) 3901–3903.

[3.4] Dalgliesh, C.E. *J. Chem. Soc.* (1952) **47**, 3940–3942.

[3.5] Pirkle, W.H. and House, D.W. *J. Org. Chem.* (1979) **44**, No.12, p1957.

[3.6] Allenmark, S.G. and Andersson, S. *J. Chromatography A.* (1994) **666**, 167–179.

[3.7] Hermansson, J., Hermansson, I. and Nordin, J. *J. Chromatography A.* (1993) **631**, 79–90.

[3.8] Bressolle, F., Audran, M., Pham, T.N. and Vallon, J.J. *J. Chromatography B Biomedical Applications.* (1996) **687**, 303–336.

[3.9] Han, S.M. *Biomedical Chromatography.* **11**, 259–271.

[3.10] Armstrong, D.W., Bagwill, C., Chen, S., Chen, J.R., Tang, Y. and Zhou, Y. *Analytical Chemistry.* (1994) **66**, 1473–1484.

[3.11] Shibukawa, A. and Nakagawa, T. *Chiral Separations by HPLC.* Krstulovic, A.M., Ellis Horwood Ltd: New York (1989).

[3.12] Moller, P., Sanchez, D., Allenmark, S. and Andersson, S. *Chiral Adsorbents and Preparation thereof as well as*

Compounds on which the Adsorbents are Based and Preparation of these Compounds. Patent SE 9203646, 1993-04-24.

[3.13] Y. Okamato, *Chem. Lett.* (1986) 1237.

[3.14] Levin, S. and Abu-Lafi, S. *"The Role of Enantioselective Liquid Chromatographic Separations Using Chiral Stationary Phases in Pharmaceutical Analysis", in Advances in Chromatography.* E. Grushka and P. R. Brown, Eds., Marcel Dekker Inc.: NY, (1993) **33**, 233–266.

[3.15] Nicoud, R-M. *Separation Science and Technology*, (2000) **2**, 476–509.

[3.16] Lowe, C.R. and Dean, P.D.G. *Affinity Chromatography.* John Wiley & Sons Inc. 1989.

CHAPTER 4

[4.1] Couillard, F. *Chromatography Apparatus*, US Patent 4,597,866, 1986-07-01.

[4.2] Rausch, C.W., Tuvin, Y. and Neue, U.D. *Improvements in chromatography*, Patent GB2110951, 1983-06-29.

[4.3] Ritacco, R.P. and Hampton, T.W. *HPLC column and column packing method*, Patent DK46188, 1988-01-29.

CHAPTER 5

[5.1] Wellings, D.A. *Preparative HPLC of Peptides*. IBC Symposium on Industrial Scale Peptide Synthesis, Las Vegas, March 1997.

[5.2] van Deemter, J.J., Zuiderweg, F.J. and Klinkenberg, A. *Chem. Eng. Sci.*, (1956) **5**, 271–289.

CHAPTER 6

[6.1] Rules and Guidance for Pharmaceutical Manufacturers and Distributors (known as the *Orange Guide*), Medicines and Healthcare Products Regulatory Agency, published by the Stationery Office, 2002.

Index

Index

Index

Note: Italicized page numbers denote entries from tables or figures.

A

B

C

D

E

F

G

H

I

L

M

N

O

P

Q

R

S

T

U

V

Printed and bound by CPI Group (UK) Ltd, Croydon, CR0 4YY
08/05/2025
01864835-0001